NUCLEAR IMAGING
IN ONCOLOGY

Nuclear Imaging in Oncology

E. Edmund Kim, M.D.

Professor and Chief of Nuclear Medicine Division
Department of Radiology
University of Texas Medical School in Houston
and
Professor of Medicine, Department of Internal Medicine
University of Texas Cancer Center
M.D. Anderson Hospital
Houston, Texas

Thomas P. Haynie, M.D.

Professor and Chairman, Department of Internal Medicine
Chief, Section of Nuclear Medicine
The University of Texas System Cancer Center
M.D. Anderson Hospital and Tumor Institute
Houston, Texas

APPLETON-CENTURY-CROFTS/Norwalk, Connecticut

0-8385-6973-0

Copyright © 1984 by Appleton-Century-Crofts
A Publishing Division of Prentice-Hall, Inc.

84 85 86 87 88 / 10 9 8 7 6 5 4 3 2 1

Prentice-Hall International, Inc., London
Prentice-Hall of Australia, Pty. Ltd., Sydney
Prentice-Hall Canada, Inc.
Prentice-Hall of India Private Limited, New Delhi
Prentice-Hall of Japan, Inc., Tokyo
Prentice-Hall of Southeast Asia (Pte.) Ltd., Singapore
Whitehall Books Ltd., Wellington, New Zealand
Editora Prentice-Hall do Brasil Ltda., Rio de Janeiro

Library of Congress Cataloging in Publication Data
Kim, E. Edmund.
    Nuclear imaging in oncology.

    Includes index.
    1. Cancer—Diagnosis. 2. Nuclear magnetic resonance—
Diagnostic use. 3. Imaging systems in medicine.
I. Haynie, Thomas P. II. Title. [DNLM: 1. Neoplasms—
Radionuclide imaging. QZ 241 K49n]
RC270.K54 1984   616.99'407575   83-25848
ISBN 0-8385-6973-0

Design: Jean M. Sabato

*To our teachers
and our children*

Contents

Preface

According to an American Cancer Society forecast for 1983, about 855,000 people will be diagnosed as having cancer excluding carcinoma in situ and nonmelanoma skin cancers; and roughly 440,000 will die of cancer in 1983. These statistics mean that more Americans than ever are facing cancer in their lifetime, and suggests a growing epidemic. Cancer remains a dreaded disease, although one-third of all cancer patients are alive at least five years after treatment. Despite modern advances, many types of cancer still progress unrelentingly, especially if discovered too late. Cancer is a formidable adversary, since a cure depends on destroying not a foreign invader, but the body's own cells, albeit transformed in character. The most dangerous tumor cells are those capable of escaping from the original tumor and forming new tumors at other sites in the body, and our knowledge of how metastasis occurs is just beginning to provide insight into basic mechanisms.

Medical imaging has become exceedingly complex. In the last 30 years, and in particular during the last decade, growth in imaging technology and in the range of techniques available has been rapid. The 1950s saw the advent of nuclear medicine and ultrasonography, and in the 1970s, computed tomography (CT), positron emission tomography, digital radiography, and nuclear magnetic resonance were introduced. This increasing sophistication of medical imaging presents a dilemma for physicians.

When CT was first introduced, it was believed that it would replace nuclear medicine imaging in many clinical situations, and to some extent this has occurred, particularly for brain imaging. However, nuclear medicine has remained useful for demonstrating tumors either indirectly, as in conventional bone and liver studies, or directly, using tumor-seeking radiopharmaceuticals. It provides a simple, safe, and noninvasive method for evaluating extent of primary tumor, presence of metastasis, and response to therapy. It can also provide valuable assistance in staging, and in determining the optimal mode of therapy. Increased patient survival as a result of improved therapy makes tumor imaging especially useful as a noninvasive method for serial tumor evaluation.

Concern has grown about the need for an organized approach to selection of imaging techniques as each new modality of medical imaging has gained acceptance. This book is designed to provide help in the selection of an optimal imaging approach to a particular cancer patient by presenting nuclear imaging studies for each category of cancer. We recognize that this information must be integrated with other knowledge in order to limit testing, and we have tried to provide at least a rudimentary discussion directed towards this end.

There are many areas of investigation using radionuclides that will prove valuable for routine clinical application. To mention only two, tracers for the various metabolic pathways in diseases and therapy may be better defined by emission computed tomography, and a most promising area of investigation for the better diagnosis of disease and of therapy is the application of radiolabeled antibodies to tumor-associated antigens. We are confident that the continuous efforts of many investigators will result in the development of even more specific technology which will extend the rewards of nuclear imaging to greater numbers of cancer patients and eventually will help in the conquest of this disease.

Acknowledgments

We are greatly indebted to our clinical colleagues at the University of Texas System Cancer Center, M. D. Anderson Hospital, and Tumor Institute for their criticism and helpful advice with this book. Dr. Howard J. Glenn reviewed Chapter 1 (Introduction of Basic Principles), and Chapter 12 (New Modalities for Nuclear Imaging); Dr. Milam E. Leavens, Chapter 2 (Tumors of the Central Nervous System); Dr. Helmuth Goepfert, Chapter 3 (Head and Neck Tumors); Dr. Clifton Mountain, Chapter 4 (Lung and Mediastinal Tumors); Dr. George Blumenschein, Chapter 5 (Breast Cancer); Dr. Josef Korinek, Chapter 6 (Tumors of the Digestive Tract); Dr. Douglas Johnson, Chapter 7 (Genitourinary Cancer); Dr. Felix N. Rutledge, Chapter 8 (Gynecologic Neoplasms); Dr. Naquib Samaan, Chapter 10 (Endocrine Tumors); and Dr. Raymond Alexanian, Chapter 11 (Hematologic Tumors). Additionally, our many thanks to Catherine Kenig, Linda Watts and Betty Hornung for their secretarial assistance in preparing the book, Dr. Anthony G. Bledin, and Dr. Carlos R. Gutierrez for their help in choosing illustrations as well as the many authors and publishers who allowed us to use their previously published case illustrations. It is a pleasure to acknowledge the kindness that we received from the editor of the series on the *Current Practice in Nuclear Medicine*, Dr. Sheldon Baum, and Appleton-Century-Crofts.

E. Edmund Kim, M.D.
Thomas P. Haynie, M.D.

NUCLEAR IMAGING
IN ONCOLOGY

CHAPTER 1

Basic Principles

PRINCIPLES OF CANCER DIAGNOSIS AND TREATMENT

Cancer has been the second leading cause of death in the United States for the past few decades. Prevention, detection, and treatment are the three major approaches to slowing the cancer problem. Obviously, the ideal solution is prevention. One might accomplish this either by eliminating exposure to carcinogens or by fortifying resistance of the host to cancer. This is not very realistic because of all the possible environmental factors and the lack of a common etiologic agent with which to vaccinate the host against all possible cancers.

Early detection of cancer, before the invasive phase, holds more promise than does prevention. Early detection depends on educating the laity and physicians. The early signs of cancer in different sites have been stressed, but fear of the diagnosis is the element that delays the patient's seeking advice. Only through constant awareness will the asymptomatic patient with occult localized disease be found in a stage in which therapy can most often prove successful. At some sites, effective screening is possible—e.g., cytology for cervix cancer. Screening procedures such as gastric and colonic fiber-optic endoscopy are highly effective for early detection of stomach and colon cancer; however, they are too costly to utilize in other than specific high risk groups. Once the diagnosis is suspected, a careful work-up is essential and proper consultation is necessary. Surgical biopsy or excision is the single most important procedure in establishing a firm diagnosis, since tumors can masquerade as benign or inflammatory conditions. Whenever possible, a histologic diagnosis is essential before undertaking radical treatment. Once the pathologic diagnosis is made, certain additional tests are advised to determine the anatomic extent of the process. Noninvasive radionuclide imaging, ultrasonography, and computed tomography are playing an important role in detection of primary and metastatic lesions. Selective angiography often allows for preoperative diagnosis.

It is important to distinguish staging from classification of cancer. Staging is an attempt to define the true extent of cancer in its three compartments (T, N, M) at a point in time, usually at detection. It does not imply a regular and predictable progression. It is arbitrary and its effectiveness is determined by whether or not a consensus exists to use it as standard terminology for treatment selection and end-stage reporting. The following is a typical type of stage grouping:

> *Stage I* (T_1, N_0, M_0): A mass limited to organ or origin is operable and resectable with only local involvement. There is absence of nodal and vascular spread.
>
> *Stage II* (T_1 or T_2, N_1, M_0): There is evidence of local spread into surrounding tissue and first-station lymph nodes. The specimen shows evidence of microinvasion into capsule and lymphatics. The lesion is operable and resectable, but there is uncertainty as to completeness of removal.
>
> *Stage III* (T_3 or T_4, N_2 or N_3, M_0): Examination reveals an extensive lesion with fixation to deeper structure, bone, and nodal invasion. The lesion is operable, but not resectable.
>
> *Stage IV* (M_1): There is evidence of distant metastases beyond local site of origin, and the lesion is inoperable.

The complexity of treatment in a patient with malignant disease demands that therapy be undertaken only by those who are capable and properly trained. The basic principle in therapy is to cure the patient with minimal functional and structural impairment. The decision as to how aggressive treatment is to be is determined by a combination of the following factors: (1) the aggressiveness of the cancer, (2) the predictability in regard to its spread, (3) the morbidity and mortality rate of the therapeutic procedure, and (4) the cure rate for the therapeutic procedure. If cure is virtually impossible, one must be guided by palliation. Although there is general agreement on many principles of management, much controversy exists over different procedures that offer limited success. Teamwork between a medical oncologist, surgeon, radiotherapist, and pathologist is required for the best effort. Maturity of decision rests on clinical experience. Optimal treatment is usually designed to treat the defined overt cancer and suspected occult deposits. Lacking such precision, the selection of treatment should be on a multidisciplinary basis, since the evidence in many cancers indicates that combined treatments are more effective than single modalities in the majority of instances. The basis of the multidisciplinary approach to cancer management is the ability of all responsible disciplines to be represented in the initial decision-making.[1] Cooperative clinical studies have been developed to answer complex questions in the treatment and reconcile different therapeutic claims.

Surgery has always played a dominant role in the diagnosis and treatment of those cancers that appear to start as a focus of malignant calls proliferating and invading the surrounding tissues. Surgeons have defined the safe limits of resections, improved the techniques of reconstruction, and learned how to operate with limited morbidity. The ideal in radiotherapy of malignant disease is achieved when the tumor is completely eradicated and the surrounding normal tissues show no evidence of structural or functional injury. In clinical practice, this ideal selective effect is obtained infrequently. Radiation therapy acts via a selective destructive effect on the tumor, not as a cautery. Chemotherapeutic agents are generally more effective for small tumor cell burdens than large, within a given histologic classification. They are generally more effective in rapidly dividing cells than in resting cells, and they may have maximal effect in further reducing tumor cell burden following noncurative surgery or radiation therapy. Chemotherapy before clinical recurrence or consolidation of a remission with continuing chemotherapy may reduce tumor cell burden to levels at which immunologic mechanisms are considered more likely to be effective. Recent attempts at immunotherapy, while based on more complete understanding of immune mechanisms, have often been disadvantaged by involving preterminal patients with very large tumor cell burdens. Already impaired immune mechanisms have little chance of producing demonstrable improvement.

CURRENT TUMOR IMAGING

There is no really good test that can rapidly screen the whole body for cancer. Tumors can be localized by radionuclide imaging techniques by two main methods. The commonest method is to use a radiopharmaceutical that localizes in normal tissue, and the tumor or other abnormality is then shown up as an area of decreased radioactivity. The alternative method is to use tumor-localizing agents that concentrate in lesions to a level higher than that of the surrounding tissue. Other techniques such as ultrasonography, computerized tomography, and x-ray contrast studies are also used. The choice between the various techniques is difficult and confusing. However, radionuclide techniques generally are noninvasive, give little radiation hazard, and yield results with an accuracy that is not too different from other tests. Thus, radionuclide studies are often first in the series of techniques used in the investigation of patients with suspected tumors.

Increased application of sophisticated imaging systems has moved the specialty of nuclear medicine in a short time into the age of enhanced accountability. Emphasis on cost containment is stimulating us to determine the least amount of diagnostic imaging that is clinically applicable and necessary.[2]

Nuclear Instrumentation
The accuracy of a particular test depends critically on the technique and the apparatus used. Rectilinear scanners are often used to image the whole body continuously from head to feet. They can deal with a wider range of radionuclide energies. Gamma cameras also produce a whole body image, but these tend to work better with low-energy isotopes. At these energies the gamma camera resolution falls off rapidly with depth. Recently, emission tomographic scanners have become available, and reproduce the variation in radioactivity throughout the body in the form of either a longitudinal or

transverse section. The advantage of this approach is the ability to see activity at depth separate from the radioactivity in superficial structures. To some extent it improves the diagnostic accuracy, especially when the information about depth helps in making a differential diagnosis. Nuclear medicine image quality is determined by contrast resolution (lesion to background counts), and spatial resolution and sensitivity of the instrumentation.[3]

Radiopharmaceuticals for Negative Tumor Localization

Most normal tissues are imaged with Tc-99m labeled agents because of their favorable physical properties. Tc-99m sulfur colloid is used to image the reticuloendothelial system in the liver. Although tumors are said to have some reticuloendothelial cells, generally they do not take up colloidal preparations to any extent. Tumors of the spleen can also be imaged using Tc-99m colloid or denatured red blood cells (RBC).

It is possible to image the pancreas with the use of Se-75 selenomethionine. Pancreatic tumors are seen as nonfunctioning areas within the normal uptake in the pancreas. Various attempts have been made to improve the situation by using amino acids labeled with F-18 or C-11. These have only been partially successful to date.

Radiopharmaceuticals for Positive Tumor Localization

Great efforts have been made and are being made to devise a radiopharmaceutical that localizes in cancerous cells. Once the radiopharmaceutical is in the region of tumor, the ability of the tumor to concentrate it depends upon the membrane transport and intracellular fixation. Apart from the use of I-131 for thyroid tumors, the limited use of Se-75 selenomethionine for liver tumors, and the use of labeled chloroquine analogues in melanomas, labeled substrates have not yet proved to be useful for tumor imaging.

Tc-99m phosphates are used extensively in bone tumor imaging due to its chemiadsorption to the hydroxy apatite of bony crystal. Tc-99m pertechnetate or DTPA shows up brain tumors because of a localized increased permeability of the blood–brain barrier.

Suggestions for general tumor scanning agents have so far centered mainly on the idea of using labeled molecules that might be included in amino acids, and nucleosides that would be involved in the increased metabolism of the cancerous cell. There is a vast literature on the use of tumor-localizing agents in virtually every clinical situation. The poor uptake of most of the agents and the low activities usually employed reduce the chance of their ever being used as efficient early diagnostic tests. The majority of the work has centered on the use of Ga-67, probably because it is generally available throughout the world.

    a. Ga-67 citrate. Ga-67 seems to be most valuable in detecting occult malignant lymphoma lesions in the mediastinum, and in differentiating hepatomas from other abnormalities in patients with liver cirrhosis. Cerebral infarcts in general do not show any accumulation of Ga-67 despite clear uptake of Tc-99m. In the chest, Ga-67 has usefully delineated the extent of bronchial carcinomas. It appears that Ga-67 is useful in the follow-up of the testicular seminoma in situations where CT scanning is not possible or is not thought to be worthwhile.

    b. Radiolabeled Antibiotics. High success rates have been found in tumors of the lung, pancreas, esophagus, and brain when imaged with Co-57 labeled bleomycin. Consistent localization of In-111 labeled bleomycin in both primary and secondary tumors was also found with an overall accuracy of 80 percent.

    c. Radiolabeled Antibodies. Most of the work concerning labeled antibodies has concentrated on those against caxcinoembryonic antigen (CEA). Using the blood background subtraction technique, successful scans have been made in the primary sites and in secondaries of various tumors. Although the uptake levels of CEA are unimpressive when compared with those in radionuclide bone scanning, the antibodies are believed to be specific for malignancy—unlike Ga-67 imaging.[4]

TUMOR IMMUNOLOGY

The well documented occurrence of spontaneous regressions of human tumors has led to widespread interest in host defenses against tumors. The possibility of developing effective immunologic approaches to the therapy of neoplastic disease has been strengthened by increasing evidence for the occurrence of host immune responses to a variety of human tumors. Necessary for such responses are tumor-specific (or tumor-associated) transplantation antigens (TSTA), which have been demonstrated in several human tumors.

A tumor-specific antigen is any malignant cell structure that is absent in or on healthy cells of the same tissue at the same stage of development, and capable of inducing immunologic reactions either in the host or by inoculation in a foreign host. A tumor-associated antigen is a cell-surface or intracellular antigen that is found on normal cells only under special circumstances, or on adult normal cells, but in reduced concentrations.

During the last decade, there has been an exploration of knowledge and methodology in molecular biology, cellular biology, and clinical investigation, with particular emphasis and activity in the field of immunology. Major advances have been seen in the understanding of the immune response and its regulation.[5] This has allowed a much more precise concept of how approaches to cancer immunodiagnosis and immunotherapy should be developed. Helper and suppressor T lymphocytes regulate the expression of antibody production and cellular immunity.

Inhibition of human tumor nodule growth in vivo by mixing suspensions of white blood cells and tumor cells from a given patient has suggested a cell-mediated reaction to the tumor.[6] Tumor antibodies directed against human tumor cells or constituents have also been searched for and demonstrated in sera of patients with Burkitt's lymphoma, malignant melanoma, osteosarcoma, neuroblastoma, and digestive system carcinoma. Since it is known that suppressor cells (either T cells or macrophages) may be increased in cancer patients, and since the administration of suppressor cells to tumor-bearing animals may cause enhanced tumor growth, it seems logical that drugs that abrogate suppressor-cell activity might be useful in cancer treatment. Indomethacin and cimetidine can abolish suppressor cell activity and augment conventional antitumor therapy in animal tumor models. The mechanisms involve inhibition of prostaglandin synthesis and blockade of H-2 receptors or suppressor cells. Another area of major advance in basic immunology involves the identification and characterization of mediators through which cellular interactions of the immune response and host-defense mechanisms are carried out. Since some have been produced in pure form through genetic engineering, they become important for further immunologic study. Several of the interferons and thymic hormone preparations are already in clinical trial.[7]

Growth in vivo of tumors that possess tumor-specific transplantation antigen suggests a deficient host response to the TSTA, possibly due to specific immunologic tolerance to TSTA, suppression of immune response by carcinogens, treatment, or the tumor itself. Considerable interest attends the speculation that interactions of lymphoid cells with cells possessing different surface antigens might serve as a surveillance mechanism capable of rapidly eliminating neoplastic cells in the absence of specific immunity. The key to advances in the new cancer immunobiology, which offer a great promise for understanding the nature of the malignant process and for developing cancer immunodiagnosis and immunotherapy, are based on new methodologic developments that include improved tumor antigen purification, characterization of cell-surface antigens, receptors, and markers of normal and malignant cells, hybridoma methodology yielding monoclonal antibody, mammalian cell cloning, and recombinant DNA methodology.[8]

Oncogenic viral antigens induced by the same virus have identical antigens between different tumors, different animals within the same species, or different species. Idiotypic antigens are present in experimental tumors induced by chemical carcinogens, and are specific for a given tumor in one individual, and are not found in other tumors induced by the same carcinogen in the same species. Embryonic antigens are normally present in the embryo at some stage in its development, and are absent in the adult or are present only in trace amounts. Division antigens appear during mitosis. The antigens and antibodies can be used to characterize the tumor cell surface, to catalog its antigens, and to define their functional and structural role in the tumor cell. The availability of highly specific antibody to the tumor cell-surface antigens and the purified tumor antigens themselves allow for serologic immunodiagnosis and development of specific tumor cell vaccines.

One of the great advances in science in recent years has been the development of monoclonal antibody methodology.[9] When recently immunized lymphocytes are hybridized with continuously proliferating myeloma cells and grown in special tissue culture medium that will only allow the hybrid cells to grow, and when these cells are carried through single cell cloning, clones of cells making antibody to individual antigens of the immunizing material can be developed. In immunodiagnosis, they can be used for radioimmunoassay studies on patients' sera, or after labeling with Tc-99m or In-111, they could be used for in vivo localization of tumor nodules. There are several approaches to immunotherapy with monoclonal use of the antibody as a carrier of toxins, use as a carrier of antitumor radioisotopes such as I-131, or use as a carrier of an antitumor drug such as Adriamycin. Recombinant DNA methodology is probably the most important technologic advance of science as it relates to the new cancer immunobiology.[10] One type of human leukocyte interferon, one of the thymic hormones, and human insulin, are being produced by recombinant DNA methods and have already been entered into clinical trials.[11]

Since many malignancies are disseminated at the time of diagnosis or disseminate shortly after diagnosis, systemic rather than local treatment is mandatory. In regard to chemotherapy, there are a limited number of malignancies for which a substantial cure rate by chemotherapy exists. In contrast, biologic therapy or immunotherapy based on unique characteristics of tumor cells has barely been explored.

BIOLOGY AND PATHOLOGY OF CANCER

There have been many definitions of cancer, most of which imply that cancers are autonomous growths that are irreversible. Accumulating experience with tumor maturation and spontaneous regression forces the recognition that irreversibility is not an essential feature of cancerous states.[12] Regardless of the attempts at definition, a few basic facts regarding cancer have been recognized:

  a. Tumors can arise only from cells that have the ability to proliferate, and they are frequently multicentric within the same tissue.[13]
  b. Hyperplasia and dysplasia often precede the development of tumors by months, and a few cancers spontaneously regress.
  c. Tumor cells do not dedifferentiate, and some solid tumor cells may replicate more slowly than do coexistent normal cells. Cancer cells may lie dormant for a prolonged period.[14]
  d. Cancers may arise following a variety of stimuli (chemical, physical, or viral), but usually only after a prolonged latent period.

The difference between the malignant cell and the normal cell lies in metabolic or biochemical processes. The hypothesis provides the rationale for chemotherapy and radiotherapy. There is a higher rate of aerobic and anerobic glycolysis found in malignant cells—the transformation of glucose into lactic acid increases under varying oxygen conditions. Malignant cells show an increase in enzyme activity. Chromosomes are believed to play an important role in malignancy, since they are concerned with cell multiplication and transmission of altered cell behavior. Normal cells in isolated states in monolayer cultures will lose mobility when another normal cell is contacted. However, malignant cells have independent movement and little intercellular adhesions. Cell turnover is often slower in rapid growing tumors than in some normal tissues, possibly due to the lack of the cell removal mechanism in malignant sites. The ability of tumors to induce their own microcirculation has suggested elaboration of an angioblastic substance, and circulation and diffusion patterns in neoplasia are keys for distribution of pharmaceuticals.[15]

Numerous studies have demonstrated a high incidence of circulating cancer cells in systemic venous blood, and an even higher incidence in regional venous blood, particularly following surgical manipulation. No definite correlation has yet been shown between the presence of circulating cancer cells and either the later development of metastases or the length of patient survival. The growth of a metastasis depends on the provision of a proper internal environment. The "soil" as well as the "seed" seems to determine the distribution pattern of metastasis. Certain viscera are better growing sites for certain tumor cells. Mechanical factors such as transcapillary passage, filtration, and cancer cell size versus lumen of vessels, are important factors in determining the location and distribution of metastases. Once a metastasis is established, many local and systemic factors will influence its growth.

The exact cause of cancer remains undetermined. Although there are readily recognizable histopathologic differences between the cancer cell and the normal cell, few metabolic differences have been determined. The concern about radiation exposure as a carcinogenic agent seems out of proportion to the small role radiation has as a definitive etiologic factor, accounting on its own for a very small fraction of all cancers. With the increasing knowledge of DNA metabolism, research workers are postulating a common ground for carcinogenesis including hereditary, viral, chemical, and radiation factors.

More than 270 types of human tumors have been recognized and defined histologically. The degrees of variation, the diagnosis, and the definition of cancer almost always depend on microscopic examination, yet histologic cancer is not equivalent to clinical cancer or to the cancer state seen in the individual patient.

Correlation of the natural history of each type of clinical disease with the type and arrangements of cells as seen by light microscopy has permitted the empiric recognition, definition, and classification of cancers. The objectives of classification are to: (1) aid the clinician in the planning of treatment, (2) give some indication of prognosis, (3) assist in the evaluation of treatment results, (4) assist in the continuing investigation of cancer, and (5) facilitate the exchange of information. The essence of a meaningful classification depends on quantifying the extent of the tumor. Most new classifications attempt to define the primary site as T_1, T_2, T_3, or T_4 with increasing extent; advancing nodal disease as N_0, N_1, N_2, or N_3; and the presence or absence of metastases as M_0 or $M+$, respectively. This system allows for considering methods of malignant spread: T for primary or direct involvement, N for secondary or lymphatic involvement, and M for vascular dissemination. For cancer at accessible sites, clinical examination and radiographic procedures are able to determine the true extent of cancer with reasonable accuracy. For cancer at inaccessible sites, surgical exploration and biopsies of selected sites or of nodes may be used to evaluate the true extent.

The differentiation between the benign and malignant tumor is usually not difficult. A typical benign tumor is encapsulated by connective tissue. Degenerative changes occur much less frequently in benign tumors than in malignant tumors. However, if a benign tumor has been present long enough, has grown large enough, or has had some impairment of its blood supply, then regressive changes can take place. Microscopically, the pattern of benign tumors is orderly, and the cells are usually uniform in size, shape, and stainability. Mitotic figures may occur in fairly rapidly growing cellular benign tumors, and their presence may be evidence of malignancy. Benign tumors do not metastasize, but if they are in a strategic location, they may cause major problems from compression of vital structures. A benign tumor sometimes may develop into a malignancy.

Malignant tumors usually do not have capsules or, if a capsule is present, it is incomplete. Gross extension into surrounding tissues or evidence of involvement of blood vessels or contiguous lymph nodes may be observed. Malignant tumors frequently show areas of necrosis. Microscopically, the malignant tumor invariably has a disorderly pattern. Mitotic figures may or may not be present, and if abnormal forms are seen with asymmetrical spindles or giant forms, then the probability of malignancy is high. Microscopic search may reveal tumor within veins, lymphatics, or perineural spaces. Morphologic characteristics can often be correlated with biologic behavior of some cancers. The presence of large numbers of lymphocytes and/or plasma cells at the periphery of some malignant tumors, as well as histiocytes in the regional lymph node sinusoids, may be related to a better prognosis. These suggest the possibility of immunologic defense mechanisms.

The following classification of tumors is based on histiogenesis (tissue of origin and cell type):

Histogenesis	Benign	Malignant
1. Epithelial Tumors		
Surface	Papilloma	Carcinoma
Glandular	Adenoma	Adenocarcinoma
2. Connective Tissue		
Smooth muscle	Leiomyoma	Leiomyosarcoma
Striated muscle	Rhabdomyoma	Rhabdomyosarcoma
Fibrous tissue	Fibroma	Fibrosarcoma
Cartilage	Chondroma	Chondrosarcoma
Bone	Osteoma	Osteosarcoma
Fat	Lipoma	Liposarcoma
Blood vessels	Hemangioma	Hemangiosarcoma
Lymph vessels	Lymphangioma	Lymphangiosarcoma
3. Hematopoietic Tumors		
Lymphoid tissue		Hodgkin's disease
		Non-hodgkin's lymphoma
		Leukemia
Granulocytic tissue		Myelocytic leukemia
Erythrocytic tissue		Erythroleukemia
Plasma cells		Multiple myeloma
4. Nervous Tissue Tumors		
Glial tissue		Astrocytoma
Meninges	Meningioma	Meningeal sarcoma
Nerve cells	Glanglioneuroma	Neuroblastoma
Nerve sheaths	Neurolemmoma	Schwannoma
Retina		Retinoblastoma
Adrenal medulla	Pheochromocytoma	

5. Tumors of More Than One
   Tissue

Breast	Fibroadenoma	Cystosarcoma phylloides
Embryonic kidney		Wilm's Tumor
Uterus		Mixed mesodermal
Multipotent cells	Teratoma	Teratoma

6. Miscellaneous

Ovary	Granulosa-theca tumor	
	Brenner tumor	
	Arrhenoblastoma	Carcinoma
	Gynandroblastoma	Dysgerminoma
Testis	Intestinal cell tumor	Seminoma
		Embryonal carcinoma
		Choriocarcinoma
		Yolk-sac tumor
Placenta	Hydatiform mole	Choriocarcinoma
Thymus	Thymoma	Thymoma
Melanoblasts	Pigmented nevus	Melanoma

BASIC PRINCIPLES OF RADIATION BIOLOGY

Physical Changes

Radiation exerts its affect on chemical compounds through two mechanisms: (1) excitation and (2) ionization of the constituent atoms. Excited atoms are more reactive chemically than in the unexcited state, and the chemical bonds may be disrupted with the breakup of complex molecules. Ionized atoms become free radicals that are highly reactive and will also disrupt chemical bonds.

X-rays and alpha-rays release their energy via orbital electron collisions, and hard or high-energy x-rays can penetrate deeply into tissue. While particles as large as protons are slowed by collision, there is a maximal release of energy in a relatively short portion of the track. Electrons are slowed gradually, and a sudden release of energy occurs. Neutrons interact with atomic nuclei rather than orbital electrons and transfer their energy by elastic collisions and recoil protons. Linear energy transfer (LET) refers to the quantity of energy deposited along the path of the beam, and oxygen enhancement ratio (OER) refers to the difference in effect on oxygenated versus hypoxic cells for a given quality of beam. Relative biological effect (RBE) refers to the efficiency of one radiation beam to another, given the same biological criteria. The RBE has been shown to increase with increasing LET, peaking at 110 Kev/u and then decreasing thereafter. It is also known that the RBE varies with dose.[16] The effective tissue dose would be dependent on the RBE, depth dose, and OER, and in turn the RBE may vary with the dose delivered.

Clinical Changes

A direct hit is the result of ionization and subsequent chemical changes of biologically important molecules, and an indirect hit is the result of interaction with free radicals produced by ionization. Free radicals can be detected by electron-spin resonance and are characterized by a single unpaired or hole electron. Although damage to DNA is not essential to all cell-killing, the basic lethal action of radiation is in the damage to the DNA. Radiochemical changes in RNA protein or a lipid molecule are less likely to be effective since the body has a surplus of identical molecules. Single-strand breaks in DNA are lesions that occur commonly, but are believed to be repairable. Repair capacity is affected by the presence (or absence) of oxygen, which enhances the action of X and gamma irradiation. Oxygen increases certain free radical formation in the radiolysis of water and makes more oxidizing radicals available. Oxygen is not as critical for the action of high LET particles such as neutrons, as damage is less dependent on radiolysis. Certain substances containing an NH_2 group or causing anoxemia, and chelating agents, protect against the effects of irradiation. They all have the common characteristic of having a reducing capacity, or of acting as scavengers. The sulfhydryl compounds may be competitive with free radicals for S–S bridges in vital enzymes or in the key amino acids in DNA molecules by donation of an H atom.

Cellular Changes

The main observed cellular effects are those due to chromosomal changes. Disturbances can be produced by chromosomal breakage that result in unequal division of chromatin material between the two daughter cells. Chromosomal injuries could be due to protein synthesis inhibition, which prevents chromosomal rejoining and contributes to cell death.

Exponential cell killing refers to fractional killing—i.e., killing of a percentage of the cell population independent of cell numbers for a specific dose. The slope of multitarget response or the survival curve is referred to as D_0 or D_{37} and is a measure of cell radiosensitivity. The more radiosensitive a cell, the steeper the curve. Most mammalian cells have a D_0 value from 50 to 280 rads. A general pattern of maximum sensitivity of cells in mitosis and maximum resistance of cells in late synthesis has emerged. In cycling cells, the repair of sublethal injury occurs 2 to 4 hours after modest doses, e.g., 500 rads. Efficient delivery of small doses may be more effective than single large doses due to induced synchrony of cycling cells and reoxygenation, which induces the cells to proliferate and thus be more sensitive to radiation.

Effects on Specific Tissues

The sequence of clinical events after the initiation of radiation therapy will be considered in four successive periods: acute clinical period (first 6 months), subacute clinical period (second 6 months), chronic clinical period (second through fifth year), and late clinical period (after 5 years). Simple radionecrosis includes changes in normal tissue that has had its tolerance exceeded. Complicated radionecrosis include changes due to irradiation of a tumor that has already destroyed normal tissue by tumor invasion.

The hematopoietic system is the most sensitive of the vital tissues. If part of the hematopoietic system is shielded, the radiation tolerance is greatly incre~ ·d and then the next most sensitive system—the gastrointestinal tract—determines tolerance. The tolerance of the organism varies with dose, time, volume, and quality of the irradiation.

    a. Acute radiodermatitis or radioepithelitis denudes the epidermal layers at 3 to 6 weeks after the start of radiation therapy and recovers in a few weeks therafter. Months later, fibrosis and atrophy occur.

    b. The acute changes result in nausea from gastric irradiation. Recovery is often complete, but necrosis can result in obstruction.

    c. If kidney tolerance is exceeded, slow vascular occlusion leads to radiation nephritis. Similar changes in bladder lead to ulceration.

    d. Manifestation of radiation pneumonitis in the weeks after radiotherapy are of clinical significance only if more than 25 percent of pulmonary volume is affected.

    e. Lenticular opacities can be produced at levels of 300 to 1000 rads, and progressive cataracts can occur with doses over 2000 rads.

    f. The acute stage for central nervous system (CNS) is silent, but the subacute and chronic stages are due to a vascular obliteration simulating strokes.

    g. Irradiation of growing cartilage results in arrest and eventual shortening of bone. This is not severe until 2000 rads is delivered.

    h. Unless the whole of the bone marrow is exposed, complete recovery is the general rule.

REFERENCES

1. Rubin P: A unified classification of cancers: An oncotaxonomy with symbols. Cancer 1973; 1:963–982
2. Potsaid MS: Diagnostic imaging in perspective. JAMA 1980; 243:2412–2417
3. Kaufman L: Nuclear medicine: Instrumentation, in Marguilis AR et al (eds): Alimentary Tract Radiology, vol 3, Abdominal Imaging. St. Louis, W. B. Mosby, 1979, pp 48–53
4. Kim EE, Deland FH, Bennett S, et al: Radioimmunodetection of cancer: An update. Allergol et Immunopathol 1980; 8:603–612
5. Bach FH, Alter BJ: Genetics of cell–cell interaction with cell mediated lympholysis, in Bach FH et al (eds): T and B Lymphocytes: Recognition and Function. New York, Academic Press, 1979, pp 527–543
6. Southern CM: Evidence for cancer-specific antigens in man. Prog Exp Tumor Res 1967; 9:1–10
7. Hersh EM: New cancer immunobiology—An overview. Cancer Bulletin 1981; 33:187–191
8. Kung PC, Goldstein G, Reinberz EL, et al: Monoclonal antibodies defining T cell surface antigens. Science 1979; 206:347–349
9. Kohler G, Milstein C: Continuous cultures of fused cells secreting antibody of predefined specificity. Nature 1975; 256:495–497
10. Burke DC: The type I human interferon gene system: Chromosomal location and control of gene expression. Interferon 1980; 2:47–64
11. Gutterman JU, Blumenschein G, Alexanian R, et al: Leukocyte interferon induced regression in human metastatic breast cancer, multiple myeloma, and malignant lymphoma. Ann Intern Med 1980; 93:399–406
12. Dewys WD: Studies correlation growth rate of a tumor and its metastases and providing evidence for tumor related systemic growth retarding factors. Cancer Res 1972; 32:374–379
13. Urban J: Bilateral breast cancer. Cancer 1969; 24:1310–1315

14. Sugarbaker E, Ketcham A, Cohen A: Studies of dormant tumor cells. Cancer 1971; 28:545–552
15. Rubin P, Casarett G: Microcirculation of tumors. Clin Radiol 1966; 17:220–225
16. Kellerer AM, Rossi HH: The theory of dual radiation action. Current Topics in Rad Res 1972; 8:85–158

CHAPTER 2

Tumors of the Central Nervous System

INTRODUCTION

For many years radionuclide brain imaging was the best noninvasive procedure for studying patients suspected of central nervous system disease. It provided an easily performed accurate means of detecting a wide range of intracranial diseases. While the study was seldom as accurate in localizing or defining disease as contrast arteriography, it was considerably better tolerated and frequently obviated the more invasive procedures. Since the introduction of computerized tomography (CT), most institutions have noted a decline in the number of nuclear brain imagings requested. In general, CT is more accurate for mass lesions, especially brainstem, parasellar, or posterior fossa tumors and low-grade gliomas. The great advantage of CT is that the delineation of anatomic structure is greater. Not only is the lesion and its extent well defined, but the relationship to surrounding structures is exquisitely detailed.

Radionuclide brain imaging has been somewhat more sensitive than CT for detecting meningoencephalitis and some cerebral infarction.[1] Where there is a low clinical suspicion of intracranial disease either modality can be used for routine screening. Such patients include those with a significant syncope, dizziness, psychologic problems, and recurrent depression. Screening of elderly patients, particularly when atrophy or communicating hydrocephalus is of clinical concern, should be done by CT, because of its ability to visualize cerebrospinal fluid (CSF) spaces.

Radionuclide cisternography has been useful in the assessment of hydrocephalus, CSF shunt patency, and CSF rhinorrhea. As the radionuclide ascends into the cerebral subarachnoid spaces, both the regional distribution and the rate of ascent and clearance provide important information about CSF pathology. Radionuclide cisternography appears to be the best diagnostic procedure to distinguish normal pressure hydrocephalus from cerebral atrophy, and can also assess patency of ventricular shunts.

The potential usefulness of substrates for glycolysis and nucleic acid synthesis as tumor imaging agents has been compared, and it is suggested that positron-labeled analogues of thymidine, uridine, and deoxyglucose will be useful for tumor imaging by positron-emission tomography.[2] For a variety of reasons, initial emphasis in clinical nuclear magnetic resonance (NMR) imaging has focused on the brain. A high level of gray–white matter discrimination is available, and when this is coupled with multiplanar imaging, the correspondence of NMR images with anatomic sections is striking. It appears that NMR imaging will come to occupy a major role in neurodiagnosis comparable with that of CT.

CHARACTERISTICS OF CNS TUMORS

Epidemiology

Brain tumors constitute 1.7 percent of all cancer in the United States; the annual age-adjusted incidence rate is the same for all races and both sexes. Viruses have been used to produce experimental brain tumors.[3] Trauma has been associated with the development of a few meningiomas,[4] and an unusually high proportion of primary malignant lymphomas have been observed in patients subject to immunosuppression for organ transplants.[5]

Pathology and Metastasis

Tumors of the central nervous system can be classified as follows:

    a. Tumors of neuroglial cells: astrocytoma, oligodendroglioma, and ependymoma.

    b. Tumors of neuronal cells: medulloblastoma, neuroblastoma, and ganglioblastoma.

    c. Tumors of mesodermal tissue: meningioma and sarcoma.

    d. Tumors of cranial and spinal nerve roots: schwannoma and neurofibroma.

    e. Tumors of lymphoreticular system: lymphoma, histiocytosis X, and plasmacytoma.

    f. Tumors of blood vessels: hemangioblastoma and angioma.

    g. Tumors of choroid plexus: papilloma and colloid cyst.

    h. Tumors of pineal region: germinoma, pineocytoma, and cyst.

    i. Secondary tumors: leukemia, metastatic carcinoma, and sarcoma.

The astrocytomas, glioblastomas, oligodendrogliomas, and malignant lymphomas occur most often in the cerebral hemispheres; glioblastoma has a predilection for the frontal lobe and astrocytoma in children for the cerebellum. Hemangioblastoma and medulloblastoma are most often found in the cerebellum. Ependymoma and choroid plexus papillomas arise predominantly from the fourth ventricle. Craniopharyngioma arises in the hypophyseal region, and chordomas arise in the clivus and sacrum. Optic gliomas grow in the optic chiasm and optic nerves.

Ependymomas are located in decreasing order of frequency in the fourth ventricle, in the supratentorial ventricles and adjacent white matter, in the spinal cord, or arising from the filum terminale. Schwannomas and meningiomas are the two most common primary intradural, extramedullary tumors in the spinal canal. For some unknown reason meningiomas occur rarely below the conus medullaris.

Metastases outside the CNS of primary brain tumors are an extreme rarity; however, medulloblastomas may metastasize to bone, lymph nodes, and liver, while glioblastomas and ependymomas may go to lymph nodes, lung, and bone.[6,7] Spread of intracranial tumors through artificial shunts has been reported.[7] Tumors of the brain and spinal cord may seed through the cerebrospinal fluid; medulloblastoma and ependymoma are known to establish distant meningeal implants in this fashion.[8] Pineal germinoma may metastasize to lung, lymph nodes, and bone.[9]

Clinical Evolution

Symptoms and signs of brain tumors may be those of increase in intracranial pressure due to the mass of the tumor and surrounding brain edema or obstruction of the ventricles or CSF pathways. In addition, focal symptoms and signs may occur due to local pressure or irritations of the tumor on the brain or cranial nerves. Headache is the first symptom in over 20 percent of the patients. Nausea and vomiting may be the first symptom or part of the symptomatology of most intracranial tumors. Vertigo and lightheadiness are not frequent symptoms except in subtentorial tumors.

Papilledema is a common sight that is present in about 75 percent of all patients. In children, a hydrocephalic head enlargement is sometimes the first observed sign, with bulging of the fontanels and widening of cranial sutures.

NUCLEAR IMAGING

Imaging of Intracranial Tumors

With newer gamma cameras with improved spatial resolution, tumors of 1 cm or greater can usually be visualized. Smaller tumors are detected if associated hemorrhage or reactive edema occurs. The use of Tc-99m DTPA or glucoheptonate and routine radionuclide angiography also improves the diagnostic accuracy.[10] Single photon emission computerized tomography (SPECT) (Fig. 1) has shown greater sensitivity and specificity than planar scintigraphy.[11] The detectability of a tumor is related to its histologic type and its location. Table 1 shows the reported incidence of various types of brain tumor, and the true positive rate of radionuclide brain imaging from several studies.[12]

Glioblastomas represent the extreme of malignancy, and the uptake of radioactivity is generally intense. They often spread through the corpus callosum into the opposite cerebral hemisphere. They also may outgrow their vascular supply, resulting in central necrosis producing a doughnut sign. The radionuclide angiogram may demonstrate increased vascularity.

Astrocytomas are less malignant, and grade I and II astrocytomas may give negative brain scans. The only findings may be a ventricular shift by CT or angiography with little or no vascular abnormality.

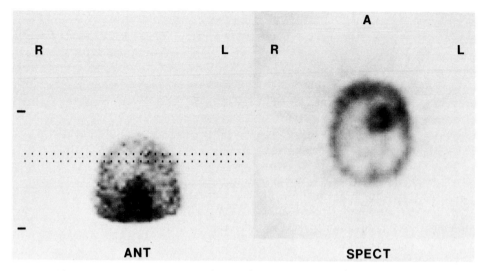

Figure 1(A). Anterior view of static brain imaging with Tc-99m DTPA shows a large lesion in the left frontal lobe. Dotted lines indicate a cursor placed at the level of the lesion. Cross-sectional image of SPECT study demonstrates clearly a doughnut-shaped lesion. Biopsy: A metastatic lesion of lung carcinoma.

Meningiomas (Fig. 2) are most commonly located parasagitally, along the sphenoid wings, on the convexity of the cerebral hemispheres, and on the floor of the anterior or posterior fossa. They are often quite vascular and show a vascular blush on the radionuclide angiogram.

The tumors that most frequently metastasize to the brain arise in the lung breast, and kidney, or from melanomas (Fig. 3). Most often the brain lesions are multiple, and they are usually supratentorial, situated subcortically in the distribution of the middle cerebral artery (Fig. 4). Most patients with metastases demonstrable by brain imaging have neurologic symptoms.[13]

Pituitary adenomas and craniopharyngiomas concentrate radiopharmaceuticals quite actively, and appear as a midline sellar and suprasellar mass.

TABLE 1.

Type of Brain Tumor	Incidence	True Positive Rate (%)
Glioma	40–45	81
glioblastoma	13–17	93
astrocytoma	7–9	73
oligodendroglioma	6–8	93
medulloblastoma	4–5	63
ependymoma	4–5	72
Meningioma	15–18	96
Metastasis	10–12	87
Pituitary adenoma	10–12	72
Acoustic neurinoma	6–7	66
Craniopharyngioma	3–4	35

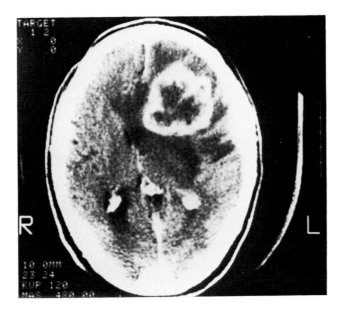

Figure 1(B). CT also shows a contrast-enhanced metastatic lesion in the left frontal lobe.

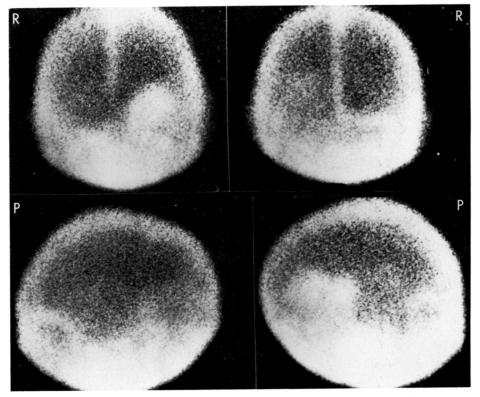

Figure 2(A). Routine four static images of brain scan with Tc-99m DTPA shows a large lesion in the left frontal lobe inferiorly. Biopsy: Meningioma.

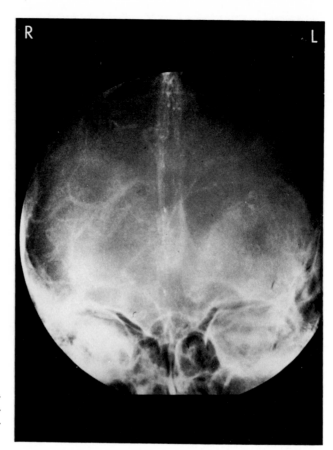

Figure 2(B). Cerebral angiogram also demonstrates a blush in the meningioma.

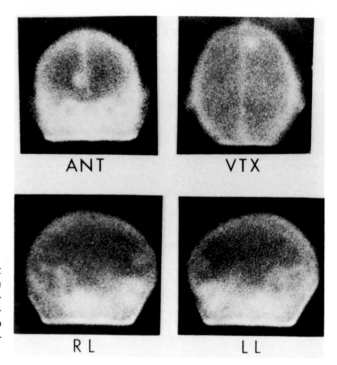

ANT

VTX

R L

L L

Figure 3. Multiple static images of brain scan with Tc-99m DTPA shows a focal lesion in the right frontal lobe anteriorly close to the midline. Biopsy: Metastatic melanoma.

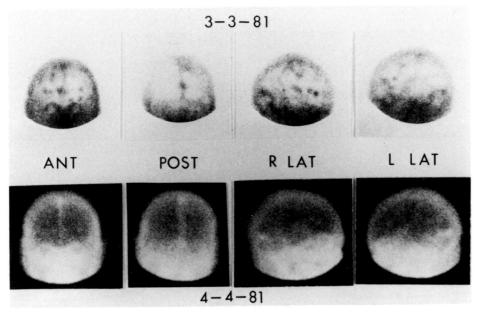

Figure 4. Routine static views of brain imaging with Tc-99m DTPA demonstrate multiple metastatic lesions of lung carcinoma in both hemispheres (upper row). Follow-up study following chemotherapy reveals some improvement (lower row).

However, they are often small, and the high activity normally at the base of the brain makes them difficult to detect.

Acoustic neuromas are usually demonstrable by brain imaging when they exceed 1.5 cm in size. The scan typically demonstrates cerebellopontine mass in the posterior view, and increased uptake at the junction of the lateral sinus and auditory meatus in the lateral view.

The use of corticosteroid to reduce cerebral edema may affect the detectability of a lesion by brain scan with Tc-99m glucoheptonate or Ga-67 citrate.[14] Presumably interstitial fluid around the lesion is reduced, resulting in decreased radionuclide concentration. The effect of irradiation therapy on serial radionuclide brain imaging for cerebral metastases has been recognized, and the decrease in Tc-99m pertechnetate uptake may be related to gliosis as well as to changes in the fine structure of the tumor and the adjacent brain tissue.[15]

The Gallium brain scan is an adjunctive procedure to be utilized in cases in which the routine radionuclide brain imaging or CT is inconclusive. In selected cases, the use of Ga-67 citrate has been helpful in the differentiation of CNS lesions as well as in the early detection of intracerebral infection.[16] It also has been suggested that Ga-67 scanning may be an important prognostic indicator in neuroblastoma, with the mean survival of 15 months for Gallium-positive patients.[17]

A positivity of 92.1 percent was obtained with Co-57 bleomycin in 80 patients with intracranial lesions, while brain imaging with In-113m produced positive results in 52 percent.[18]

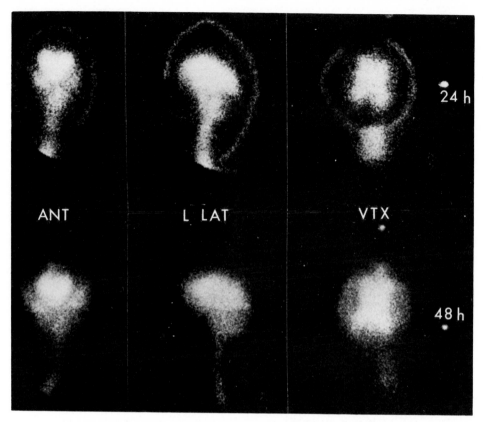

Figure 5. Anterior, left lateral, and vertex views of the head at 24 and 48 hours following the injection of In-111 DTPA in a dementic patient with lung cancer show persistent radioactivity refluxed into lateral ventricles and no migration of the activity over the convexity beyond the Sylvian cistern, indicating the normal pressure hydrocephalus.

Tl-201 seems to promise greater diagnostic precision for cerebral metastasis, with an examination time of only 30 minutes. All 25 patients with one or several brain metastases gave a positive result within 10 minutes of injection.[19] The definition of the lesion was better in 88 percent, and in 20 percent small metastases that were invisible with Tc-99m pertechnetate could be demonstrated with Tl-201.

Tc-99m phosphate bone scan agents have been used to aid in the differential diagnosis of lesions detected in routine brain images, since they accumulate more in calvarial lesions than do brain scan agents.[20,21]

Radionuclide cisternography has also been employed in cerebral tumors (Fig. 5). Twenty-four patients with supratentorial tumors were examined by the radionuclide cisternogram, and all 24 showed either partial or total subarachnoid blockage of CSF flow.[22] Most patients had multiple sites of blockage, the most common combination being the Sylvian fissure block and a unilateral convexity block. It was also found that severe subarachnoid obstruction of CSF flow, leading to impaired CSF absorption at the superior sagittal sinus, was the main cause of raised CSF pressure and papilledema

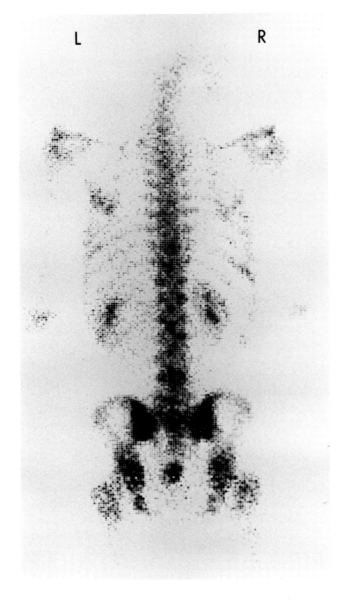

L   R

Figure 6. Posterior static image of axial skeleton with Tc-99m MDP shows multiple metastatic lesions of medulloblastoma in the lumbar and lower thoracic vertebrae.

associated with supratentorial tumors studied by cisternography.[23] A case of acoustic neuroma was identified by radionuclide cisternography while brain imaging, electroencephalogram (EEG), radiography, and CT failed to demonstrate the expected tumor.[24] It showed a well-defined round area of decreased radioactivity in the region of the cerebellopontine angle.

Radionuclide Imaging of Extracranial Metastases
Radionuclide (Fig. 6) bone and liver-spleen imaging occasionally demonstrate clinically suspected metastatic lesions of medulloblastoma, ependymoma, and neuroblastoma.

OTHER NUCLEAR STUDIES

Monitoring Therapeutic Response or Complication
Radionuclide cerebral angiogram is helpful in determining any significant spasm following surgery or contrast angiography, and is also useful in evaluation of the gelfoam embolization of intracranial meningiomas.[25] Chemotherapeutic neurotoxicity has been shown as patchy abnormal activities on brain imaging using Tc-99m glucoheptonate,[26] while CT was normal. After chemotherapy was discontinued, clinical symptoms improved and a subsequent radionuclide imaging was normal.

Monitoring Blood–Brain Barrier (BBB) Disruption
Although the passage of many materials across the capillary interface is restricted in comparison to free diffusion, this limited permeability of blood vessels is more pronounced in the CNS than in other parts. Most investigators agree with the concept that the "barrier" is a complex of morphologic and metabolic compartments.

Considerable evidence supports the view that the existence of the BBB is a principal reason for the poor chemotherapeutic responses seen in CNS malignant tumors.[27] The technologic criteria that document BBB disruption can be displayed by radionuclide brain imaging or CT with contrast enhancement.

In many cerebral neoplasms the microvascular structure is quite different from that of the normal brain. Observed changes have included hyperplastic or thinned endothelial cells with long, irregular intercellular junctions, increased number of cytoplasmic pinocytic vesicles, irregularity of the basement membrane, and absence of glial foot processes.[28] The abnormal intercellular junctions allow free passage of protein tracers, and this most likely is the major factor responsible for the abnormal brain scan in malignant cerebral tumors.[29]

Ga-68 EDTA has been used for the evaluation of BBB by positron emission transaxial tomography (PET) scanner, and Ga-67 EDTA should produce similar tomographic results using SPECT as PET.[30] I-123 iodoamphetamine (IMP) has been found to cross the BBB with a distribution proportional to blood flow[31] by SPECT (Fig. 7).

Recently, PET of the brain with Rb-82 obtained from a portable generator has been used to evaluate the integrity of the BBB in 8 patients with brain tumors,[32] and it appears a promising method since a high tissue-to-blood ratio of the Rb-82 is achieved in regions of BBB disruption.

Positron Emission Transaxial Tomography
A new era has begun in diagnostic imaging with the use of positron emission tomography for the noninvasive study of biologically important substrates. Advances in instrumentation have made it possible to obtain whole-body tomographic sections, so that a map of quantitative tracer concentration within an organ or tissue can be obtained with high resolution and accuracy. Using computerized data processing, one can measure regional physiologic functions in tomographic sections of tissue. Tumors frequently exhibit metabolic patterns that deviate from those of the tissue from which they are derived.

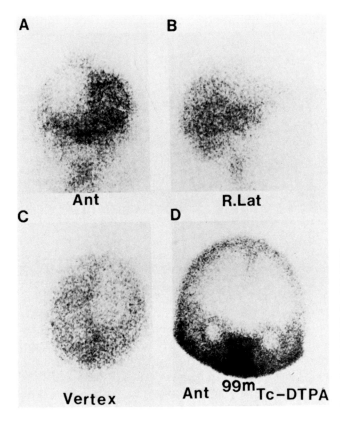

Figure 7. Anterior, right lateral, and vertex views of brain imaging with I-123 iodoamphetamine show strikingly decreased radioactivity in the area of grade III astrocytoma. Anterior image with Tc-99m DTPA fails to demonstrate a tumor. *(From J Nucl Med 1981; 22:1081, with permission.)*

Characteristically, they exhibit accelerated glycolysis and nucleic acid synthesis, a metabolic pattern that has been called the "biochemical phenotype of malignancy," which may be exploited to provide improved tumor imaging methods based on PET (Fig. 8).[33] Active uptake of F-18 2-fluoro-2-deoxyglucose by a seminoma was found in animals,[34] and absolute uptakes of H-3 thymidine, H-3 uridine, and C-14 2-deoxyglucose in sarcomas and tumor-to-blood ratios were as high one hour after injection as comparable maximum values achieved for Ga-67 citrate after 48 hours.[35] Using PET, two cases of radiation necrosis were distinguished from the three recurrent tumors. In the two cases of radiation necrosis the rate of glucose utilization in the lesion was markedly reduced. In the recurrent gliomas, however, the glucose metabolic rate was elevated.[36]

Nuclear Magnetic Resonance Imaging

It appears that NMR imaging will come to occupy a major role in neuroradiologic diagnosis comparable with that of CT.[37] A high level of gray–white matter discrimination is available, and when this is coupled with multiplanar imaging, the correspondence of NMR images with anatomic sections is striking. With NMR, unlike x-ray CT, bone artifact is not a problem and the different sequences available allow a variety of approaches to particular clinical problems. The characteristic feature of tumors is an increase in T_1 (spin-lattice relaxation time) and T_2 (spin-spin relaxation time), producing

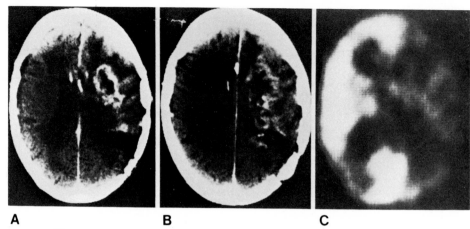

A   **B**   **C**

Figure 8. CT scans (**A** and **B**) show an enhanced lesion in the left frontoparietal lobe. F-18 FDG-PET scan (**C**) shows a hypometabolic area corresponding to the lesion. *(From Radiol 1982; 144:886, with permission.)*

low-density signals (Fig. 9). Surrounding edema shows a similar change, although this is usually minimal with benign tumors. There is frequently much more edema associated with malignant tumors. Within the tumor, discernment of more internal structure than with CT, including necrosis, cyst, and hemorrhage frequently is possible. Other nuclei such as Na-23 and P-31 may provide images of biochemical or metabolic changes in brain tumors. It is also possible to attach stable free radicals to a wide variety of other chemicals to produce a whole new range of contrast agents, as well as to make metabolites and other chemicals of biologic importance.

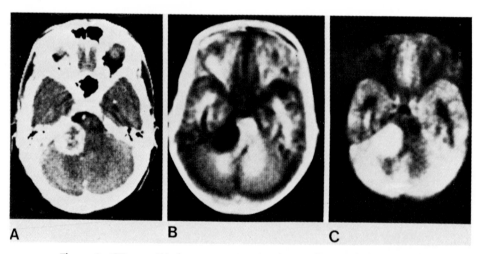

A   **B**   **C**

Figure 9. CT scan (**A**) shows a contrast-enhanced acoustic neuroma in the right cerebello-pontine angle. NMR images (**B** and **C**) show long T_1 of the neuroma, which appears dark on inversion-recovery image but light on spine-echo image. *(From Am J Neuro Radiol 1982; 3:473, with permission.)*

OTHER RADIOLOGIC METHODS

Transmission Computerized Tomography

CT is slightly better than radionuclear imaging (RI) in detecting meningiomas, glioblastomas, and metastases, but is clearly superior in detecting parasellar masses, brainstem lesions, cysts, and low-grade gliomas. CT also appears to provide more diagnostic information than RI, largely because many patients without focal signs have dementia as their primary problem. In terms of evaluating patients with nonfocal presentations for the presence of a mass lesion, RI was accurate in 291 patients (98.6 percent), while CT was accurate in 294 (99.7 percent). In paired CT–RI studies performed on 67 patients with headache as the only complaint, both tests were normal in 64 (94 percent).[38]

When focal neurologic signs are present, CT with contrast enhancement is the preferable initial examination. In these patients, RI need be obtained only when the CT is negative or does not fully explain the patient's symptoms, or when the CT is technically unsatisfactory. On the basis of currently available data, CT appears to be the superior screening test for the evaluation of the majority of patients with suspected mass lesions. However, CT is presently a more expensive study and requires a high dose of contrast media, which increased the risk of this study.

Where CT is available, it has largely replaced cisternography as a screening test for hydrocephalus, because of its exquisite delineation of the CSF spaces. However, in the case of significant ventricular delineation without atrophy, cisternography should be performed. When CT shows third and fourth ventricle enlargement, communicating hydrocephalus is strongly suspected. In these patients, cisternography is then performed to determine the need for shunting or for possible compensated hydrocephalus.

Intracranial Angiography

Angiography has been largely replaced by CT, except for the diagnosis of aneurysms, arteriovenous malformations, or vascular tumors such as hemangioblastomas. By means of angiography not only can space-occupying masses be identified but at times a specific diagnosis can be made. The angiographic findings in brain tumors depend on stretching or displacement of vessels by the mass, the demonstration of tumor vessels or a diffuse blush or stain with the lesion, early or delayed fillings of draining veins, and the visualization of an avascular area if the mass is cystic, necrotic, or has little circulation within it.

Pneumoencephalography (PEG)

PEG is particularly useful in showing small masses that may encroach upon the basal cisterns and in posterior fossa lesions. In most other situations, it has been replaced by CT.

Skull X-Ray

Skull radiographs may show signs of increased intracranial pressure, which are not specific for tumors. Frequently the shift of the calcified pineal gland only indicates the existence of a mass in the contralateral hemisphere.

Roentgen evidence of calcification is found in about 15 percent of all intracranial tumors. Meningiomas may invoke a hyperostotic response, local thinning and outward bulging, or localized erosion.

REFERENCES

1. Kim EE, Deland FH, Montebello J: Sensitivity of radionuclide brain scan and computed tomography in early detection of meningoencephalitis. Radiol 1979; 132:425–429
2. Larson SM, Grunbraum Z, Rasey JS: Position imaging feasibility studies: Selective tumor concentration of ^{3}H-thymidine, ^{3}H-uridine, and ^{14}C-2-deoxyglucose. Radiol. 1980; 134:771–773
3. Walker MD: Chemotherapy of malignant glioma. Proceedings of National Conference, 1973, vol 7, pp 817–821
4. Whatmore WJ, Hitchcock ER: Meningioma following trauma. Br J Surg 1973; 60:496–498
5. Penn I, Starzl TE: Malignant tumors arising de novo in immunosuppressed organ transplant recipients. Transplantation 1972; 14:407–417
6. Henriquez AS, Robertson DM, Marshall WJS: Primary neuroblastoma of the CNS with spontaneous extracranial metastases. J Neurosurg 1973; 38-226–231
7. Wakamatsu T, Matsuo T, Kawano S, et al: Extracranial metastasis of intracranial tumor: Review of literature and a case report. Acta Pathol Jpn 1972; 22:155–169
8. Bryan P: CSF seeding of intracranial tumors: A study of 96 cases. Clin Radiol 1974; 25:355–360
9. Borden S, Weker AL, Toch R, et al: Pineal germinoma. Am J Dis Child 1973; 126:214–216
10. Tanasescu DE, Wolfstein RS, Waxman AD: Critical evaluation of ^{99m}Tc glucoheptatonate as a brain imaging agent. Radiol 1979; 130:421–423
11. Hill TC, Lovett RD, McNeil BJ: Observations on the clinical value of emission tomography. J Nucl Med 1980; 21:613–616
12. Harbert JC, Rocha AFG: The central nervous system, in Rocha AFG, Harbert JC (eds): Textbook of Nuclear Medicine: Clinical Applications. Philadelphia, Lea & Febiger, 1979, pp 51–103
13. Delaney JF, Gertz D, Schreiner DP: Usefulness of brain scans in metastatic carcinoma of the lung. J Nucl Med 1976; 17:406–409
14. Waxman AD, Beldon JR, Richli W, et al: Steroid-induced suppression of gallium-uptake in tumors of the central nervous system. J Nucl Med 1978; 19:480–482
15. Antar M, Rembish R: The effect of irradiation therapy on serial radioisotopic brain scans for cerebral metastasis. Clin Res 1979; 27:381
16. Waxman AD: Gallium scanning in cerebral and cranial disorders. CRC Crit Rev Diag Imag 1980; 13:89–100
17. Bidani N, Moohr JW, Kirchner P, et al: Gallium scanning as prognostic indicator in neuroblastoma. J Nucl Med 1978; 19:692–696
18. Salar G, Carter A, Zampieri P: [57]Co bleomycin for brain scan diagnosis of intracranial lesions. J Nucl Med 1978; 19:137–140
19. Ancri D, Bassett JY: Diagnosis of cerebral metastasis by Thallium-201. Br J Radiol 1980; 53:443–453
20. Maeda, T, Tonami N, Nakajima K, et al: Compensative value of Tc-99m DTPA and Tc-99m MDP scintigraphy on brain lesions. Radioisotopes 1980; 29:31–33
21. Kim EE, Domstad PA, Choy YC, et al: Differential accumulation of Tc-99m DTPA

and Tc-99m pyrophosphate within cerebral and cranial lesions. J Nucl Med 1980; 21:838–840

22. Van Crevel H: Radioactive iodine human serum albumin cisternography in cerebral tumors. Neuroradiology 1979; 18:133–138

23. Van Crevel H: Papilledema, CSF pressure, and CSF flow in cerebral tumors. J Neurol Neurosurg Psych 1979; 42:493–500

24. Breitschuh E, Facorro L, Emrich D: Visualization of an acoustic neurinoma by radioisotope cisternography. Neuroradiology 1978; 15:195–197

25. Baum S, Coleman LL, Latshaw RF, et al: Radionuclide blood flow studies before and after gelfoam embolization of intracranial meningioma. Clin Nucl Med 1979; 4:412–414

26. Sherkow LH: Chemotherapeutic neurotoxicity on brain scintigraphy. Clin Nucl Med 1979; 4:439–440

27. Neuwelt EA, Diehl J, Hill S, et al: Osmotic blood–brain barrier disruption in malignant brain tumor patients: A phase I study. Proc Am Ass Cancer Res 1980; 21:365

28. Dunn JS, Wyburn SM: The anatomy of the blood–brain barrier: A review. Scottish Med J 1972; 17:21–36

29. Bar-Sella P, Front D, Hardoff R, et al: Ultrastructural basis for different pertechnetate uptake patterns by various human brain tumors. J Neurol Neurosurg 1979; 42:924–930

30. Phelps ME: Emission computed tomography. Semin Nucl Med 1977; 7:337–365

31. Winchell HS, Horst WD, Braun L, et al: N-isopropyl-(^{123}I)-p-iodo-amphetamine— single-pass brain uptake and washout: Binding the brain synaptosomes and localization in dog and monkey brain. J Nucl Med 1982; 21:947–952

32. Yen CK, Yano Y, Budinger TF, et al: Brain tumor evaluating using [82]Rb and positron emission tomography. J Nucl Med 1982; 23:532–537

33. Harap KR: Deviant metabolic patterns in malignant disease, in Ambrose EJ, Roe FJC (eds): Biology of Cancer, ed 2. London, Cox and Wyman, 1975, p 96

34. Som P, Atkins HL, Bandyopadhyay D, et al: Early detection of neoplasms with radiolabeled sugar analog. J Nucl Med 1979; 20:662

35. Larson SM, Grunbaum Z, Rasey JS: Position imaging feasibility studies: Selective tumor concentration of ^{3}H-thymidine, ^{3}H-uridine, and ^{14}C-2-deoxyglucose. Radiol 1980; 134:771–773

36. Patronas NJ, DiChiro G, Brooks RA, et al: Work in progress: [18]F fluoro-deoxyglucose and positron emission tomography in the evaluation of radiation necrosis of the brain. Radiol 1982; 144:885–889

37. Bydder GM, Steiner RE, Young IR, et al: Clinical NMR imaging of the brain: 140 cases. Am J Roent 1982; 139:216–235

38. Alderson PO, Gado MH, Siegel BA: Computerized cranial tomography and radionuclide imaging in the detection of intracranial mass lesions. Semin Nucl Med 1977; 161–174

CHAPTER 3

Head and Neck Tumors

INTRODUCTION

Radionuclide imaging has not commonly been used for evaluation of primary head and neck tumors because of nonspecific findings brought about by uptake in normal structures and the relatively small size of these lesions. Also, the low prevalence of liver and bone metastases from primary head and neck tumors limits the usefulness of scans of these organs. Physical examination, radiographic studies including CT, and biopsy have usually been used to determine the diagnosis, and for staging. Patients with choroidal melanoma associated with a secondary retinal detachment and loss of useful vision who are candidates for enucleation may benefit from Phosphorus-32 uptake study, which has occasionally been used to locate tumors at the time of planned enucleation and Co-60 implant therapy.[1]

Salivary gland imaging, while sensitive to the presence of functional alterations, is also nonspecific and thus has not been widely adopted as a diagnostic test for primary tumors. Ga-67 citrate, Se-75 selenomethiomine, and Co-57 bleomycin have been applied for imaging head and neck tumors, but have thus far demonstrated only limited usefulness.

Routine bone and liver scans in patients with head and neck carcinoma should be reserved for those patients with advanced primary tumors, with regional node metastasis, or with clinical or laboratory evidence of liver or bone involvement.[2] However, bone scans have been found to permit a better evaluation of pathological areas of bones, identify distant lesions, and provide earlier and more complete information than standard radiographs.

CHARACTERISTICS OF HEAD AND NECK TUMORS

Epidemiology

Tumors of the eye and orbit were found to comprise 0.3 percent of all malignant tumors occurring annually in the United States.[3] Malignant melanoma is the most common primary intraocular malignant tumor, although it represents only a relatively small proportion of all malignant melanomas. Intraocular melanomas are seen most frequently in patients in the sixth decade of life, and they are almost always single and uniocular. Retinoblastoma is the most common intraocular neoplasm in children. About 20 percent of the cases are bilateral, and the tumor is believed to be congenital.[4] They are most frequently diagnosed in children under 2 years of age.

Tumors of major salivary glands are rare, and they make up only 0.4 percent of all cancer.[3] The preponderance of benign salivary tumors are found in women, but malignant tumors are equally divided among the sexes. About 2 percent of these tumors occur in children.[5] The proportion of malignant tumors in three major salivary glands increases from the parotid through the submaxillary to the sublingual gland.

Malignant epithelial tumors of the maxillary antrum are found more frequently in men than in women, and occur usually in patients 40 to 70 years of age.

Carcinoma of the tongue constitutes 1.1 percent of all cancer in white men, and 0.8 percent in black men.[3] Excluding the lower lip, carcinomas of the tongue constitute the largest single group of malignant tumors of the oral cavity. Carcinoma of the tongue is often associated with poor oral hygiene. The use of tobacco, cigar and pipe smoking have also been incriminated as causative factors.

Tumors of the mandible are relatively rare, and they are generally found in relatively young patients.

Cancers of the nasopharynx were found to make up 0.2 percent of all cancer.[3] They are rare in children. The considerably greater occurrence of these tumors among Chinese has long been observed.

Cancer of the larynx constitutes about 2.5 percent of all cancer in men in the United States.[3]

Pathology and Metastasis

Malignant melanomas arise most frequently in the choroid and ciliary body.[6] All uveal melanomas are of neuroectodermal origin, and they arise from preexisting nevi in almost 100 percent of the instances. Uveal melanomas show three main cell types, and the mixed cell tumor (i.e., containing spindle B and epithelioid cells) is the most common. It is not too rare for 20 or more years to elapse before the tumor invariably becomes apparent in the liver, and fairly often in the brain. Melanomas arising in the conjunctiva may involve the cervical lymph nodes.

Retinoblastomas tend to outgrow their blood supply and undergo extensive necrosis, and they also invade the choroid with subsequent involvement of the meninges. Retinoblastomas metastasize to bones, lymph nodes, and liver in terminal stages, and may spread through the lymphatics, the blood, or the cerebrospinal fluid.[7]

In one series of 1378 benign tumors of the parotid gland, 93 percent were benign mixed tumors (pleomorphic adenomas).[8] Monomorphic adenomas include adenolymphomas (Warthin's tumor), oxyphilic, basal cell, tubular, and clear cell adenomas. A malignant transformation occurs in about 5 percent of benign mixed tumors, usually in elderly patients. Hemangiomas, lipomas, lymphangiomas, oxyphil and basal cell adenomas are rare benign tumors.

Mucoepidermoid carcinomas constitute about 20 percent of all malignant tumors of the salivary glands;[8] they arise most frequently from the parotid gland and the minor salivary glands. Adenoid cystic, acinic cell, and squamous cell carcinomas are the next most common malignant salivary gland tumors. Malignant lymphomas, hemangiopericytomas, and sarcomas are rare malignant tumors in salivary glands.

Metastases of malignant salivary gland tumors usually occur in the lymph nodes of the parotid, submaxillary, cervical, supraclavicular regions, and sometimes the mediastinal regions. The jugulodigastric and deep upper cervical lymph nodes are the most frequently involved. Metastasis to the lungs and liver may occur and bone metastases have been observed in the skull, mandible, ribs, vertebrae, and pelvis. Pulmonary metastases from a benign mixed tumor have been reported.

The majority of carcinomas of the maxillary sinus originate in the infrastructure of the maxillary antrum.[9] The overwhelming majority of carcinomas of the antrum are squamous cell carcinomas. Malignant lymphomas, melanomas, sarcomas, and plasma cell tumors may also occur. Distant metastases from carcinomas of the maxillary antrum are uncommon.

Carcinomas of the mobile portion of the tongue arise most frequently on the lateral border, and the majority of these are squamous cell carcinomas. Adenocarcinomas, adenoid cystic, and mucoepidermoid carcinomas generally are found in the base of the tongue: approximately two-thirds of all patients with carcinoma of the tongue present metastatic adenopathy at some time during the course of the disease. The nodes of the jugular chain, particularly the high subdigastric nodes, are most frequently invaded. Distant metastases are usually observed only in late stages of the disease.

Ossifying fibromas and giant cell reparative granulomas of the mandible develop most commonly in adolescents. About 70 percent of ameloblastomas are found in patients 10 to 35 years old, and Ewing's sarcomas are found generally in persons under 20 years of age. Osteosarcomas of the jaw may be found in persons of all ages.

Ewing's sarcoma quite characteristically involves other bones, the regional lymph nodes, and the lungs. Osteosarcomas metastasize with preference to the lungs, seldom to regional nodes. Burkitt's tumors metastasize to the thyroid, liver, pancreas, kidneys, and ovaries.

About 90 percent of the malignant tumors of the nasopharynx are squamous cell or undifferentiated carcinomas. Tumors of the nasopharynx metastasize frequently; over one-third may develop bilateral cervical metastases.[10] Malignant lymphomas of the nasopharynx may progress to the supraclavicular region, axilla, and mediastinum. Carcinomas may metastasize to the lungs, liver, and bones.

Clinical Evolution

Malignant melanoma arising in the choroid may produce a retinal detachment with consequent visual impairment, which is usually the first symptom. Another symptom may be pain from a secondary glaucoma caused by a forward displacement of the iris root. Leukokoria (white pupil) is the presenting sign of the majority of patients with retinoblastoma. Other signs and symptoms include strabismus, iritis, hyphemia, and nystagmus.[7]

The benign mixed tumors of the salivary glands show variable rates of growth. The first symptom in practically all patients is a small painless lump. With increase in size, there is unsightly facial asymmetry and pain. Malignant tumors of the salivary gland grow much more rapidly than do benign tumors. Paralysis of the facial nerve is present in about one-third of the patients with malignant tumors.

The majority of patients with carcinoma of the maxillary antrum first complain of a toothache or loosening of teeth. An anterolateral tumefaction is typical of most tumors of the infrastructure.

When the tongue becomes ulcerated and secondarily infected, otalgia on the same side as the lesion, a certain degree of hypersalivation, and dysphagia may occur. Difficulties with speech, dysphagia, and weight loss may follow. Pain is an important symptom in the later stages of disease.

Giant cell reparative granulomas of the mandible develop slowly and may reach a huge size. The slow progress of ameloblastomas rarely produces any symptoms. Ewing's sarcomas develop faster, and pain accompanies their growth. Osteosarcoma of the mandible is invariably accompanied by severe pain and sometimes fever.

A unilateral painless upper cervical adenopathy is often the first sign of nasopharyngeal cancer. A unilateral diminution in hearing is very commonly found accompanying tumors of the nasopharynx. Cranial nerve paralysis is not frequently the first symptom, but is common later in the development of these tumors.

RADIONUCLIDE IMAGING AND UPTAKE STUDIES

A. The salivary glands continue to present many diagnostic problems, since conventional procedures including sialography and aspiration biopsy all have drawbacks. Biopsy seems hazardous because of the proximity of the facial nerve, and the possibility of creating fistulae or seeding tumor through the biopsy tract. Tc-99m pertechnetate is actively excreted by the salivary glands much like iodide ion, and appears in the salivary glands within several seconds after injection. Unilateral disease is detected with high sensitivity. Warthin's tumor and acute inflammatory disease result in unilateral increased uptake (Fig. 1), while some primary and metastatic tumors, cysts and chronic sialoadenitis appear as unilateral decreased uptake.[11,12]

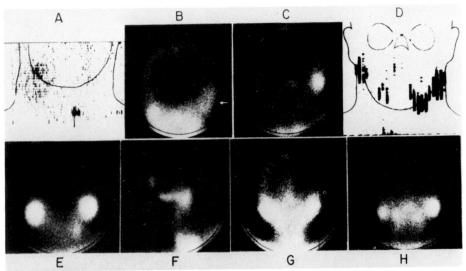

Figure 1. Salivary gland imaging with Tc-99m pertechnetate. (**A**) Anterior view, Warthin's tumor of right parotid gland (hot spot). (**B**) Posterior view, cylindroma of right parotid gland (cold defect) with metastasis to skull (arrow). (**C**) Posterior view, carcinoma of left parotid gland (cold defect). (**D**) Anterior view, right parotid carcinoma distorting contours of submandibular and parotid glands. (**E**) Posterior view, Mikulicz's disease (parotomegaly). (**F**) Left lateral view, Sjogren's syndrome (obstruction and dilatation of Stenson's duct). (**G**) Posterior view, acute parotitis of left parotid gland (increased uptake). (**H**) Posterior view, chronic parotitis of left parotid gland (diminished uptake). *(From Semin Nucl Med 1972; 2:270, with permission.)*

Although functional tumors on salivary gland imagings were thought to be benign, a case of functioning papillary adenocarcinoma of the parotid gland showing increased focal uptake of the pertechnetate has been reported.[3]

A papillary oncocytoma of the maxillary sinus was successfully imaged using salivary gland imaging, and the ability of an oncocytoma to accumulate pertechnetate, even in an ectopic salivary location suggests that the technique may be valuable in diagnosing functional oncocytomas.[14] There was a diagnosis of a vascular acinic cell carcinoma of the sublingual gland by the dynamic scintigraphy with Tc-99m pertechnetate.[15]

B. Gallium-67 citrate imaging detected 56% of primary head and neck tumors and their metastases in 65 patients.[16] The results of 1306 scans showed that a positive Ga-67 scan was associated with tumor involvement, but a negative scan cannot reliably rule out tumor involvement. Lesions over 3 cm in diameter had a significantly higher detection rate than smaller lesions, and previous irradiation or surgery did not affect diagnostic accuracy.

It has been concluded that Ga-67 scintigraphy has limited value in evaluating radiotherapy of patients with head and neck tumors.[17] Five of seven scans corresponded to a clinically proven tumor regression, while two false negatives were noted.

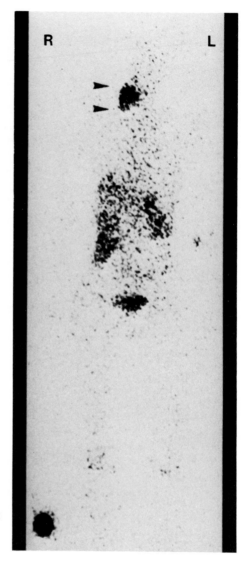

Figure 2. Anterior whole body image with Co-57 bleomycin in a patient with squamous cell carcinoma of the right epiglottis shows a prominent uptake in the right neck tumor. *(From Cancer 1979; 43:1653, with permission.)*

C. Se-75 selenomethiomine was used in 80 cases and Ga-67 citrate in 20 cases of laryngeal tumors. The results agreed with clinical and morphological results in 66 of 80 cases,[18] and also indicated that radiological and isotopic methods can correctly diagnose the type and extent of laryngeal tumors in 95 to 98 percent of the cases.

D. Co-57 bleomycin imaging (Fig. 2) appeared to be a promising technique in the evaluation of patients with head and neck tumors. Clinical and scan findings with Co-57 bleomycin concurred on the presence and extent of tumor in 9 of 11 patients (82 percent); tumor was present in 7 and absent in 2 of the 9 patients.[19]

E. In a review of 169 patients with head and neck carcinoma, the incidence of true positive bone scans (Figs. 3 to 5) was only 2 percent and no

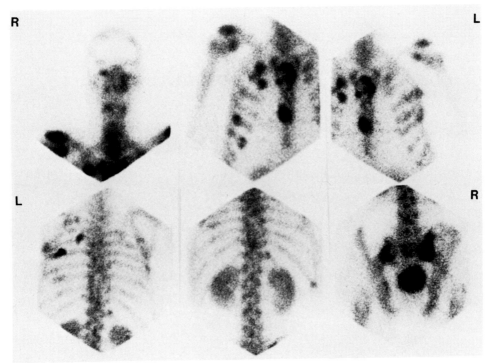

Figure 3. Routine static images of axial skeleton demonstrate multiple metastatic lesions (hot spots) of melanoma in sternum, ribs, vertebrae, and pelvis. Note a cold lesion in left iliac bone.

true positive liver scans were found.[2] However, Tc-99m bone scan is a simple and safe diagnostic technique that can demonstrate abnormality of the bone with greater sensitivity than routine radiographs. When 54 patients with malignant lesions in the mandible (Fig. 6) were examined clinically, radiographically, and scintigraphically, clinical assessment overestimated and radiography underestimated the extent of involvement of the lesions.[20] A combination of radiography and scintigraphy yielded the best information on the differentiation between an inflammatory condition and malignant tumor, and the extent of malignant growth in adjacent bone in 28 patients with malignant tumors of the nasal, paranasal, and palate regions.[21] Radionuclide liver and brain imagings can show clinically suspected metastatic lesions of head and neck cancer, especially melanoma (Figs. 7 to 11).

F. Cervical lymph node scintigraphy in head and neck tumors can demonstrate the thoroughness of lymph node removal. It may also be useful in checking the results of radiotherapy, but less so in judging the regression of metastasis.[22]

G. The P-32 uptake test has shown greater than 95 percent accuracy in differentiating benign from malignant ocular lesions.[23] The uptake was significantly lower in choroidal hemangiomas than in comparable sized melanomas.[24] It appears particularly valuable when the fundus cannot be visualized.

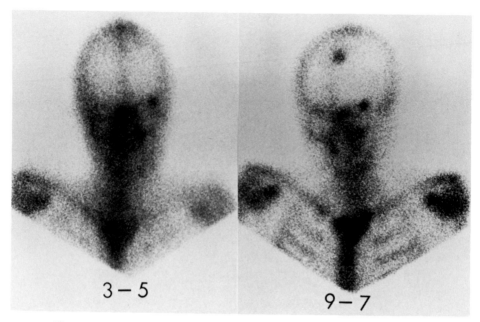

Figure 4. Anterior and left anterior oblique views of the head with Tc-99m MDP show progressive metastatic lesions in the left ortbital margin and frontal skull near the midline.

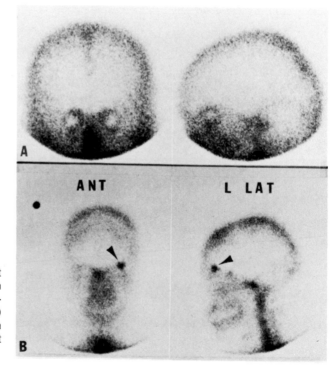

Figure 5. Frontal and left lateral views with Tc-99m DTPA (upper row) and Tc-99m MDP (lower row) show a metastatic lesion of melanoma in the left orbital margin.

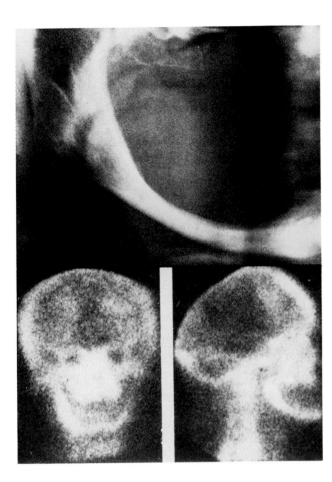

Figure 6. Radiograph of right mandible shows no definite bony destruction. Frontal and right lateral views of the face with Tc-99m MDP show increased radioactivity in the right mandible with carcinoma. *(From Acta Radiol Diag 1981; 22:485, with permission.)*

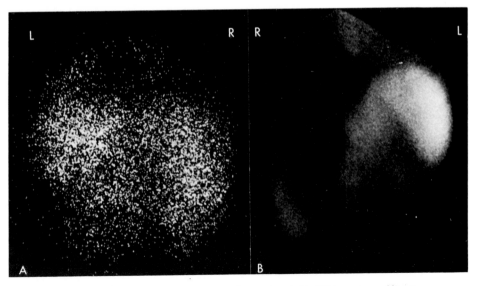

Figure 7. Posterior dynamic (**A**) and anterior static (**B**) imaging of liver-spleen with Tc-99m sulfur colloid show almost complete replacement of right hepatic lobe with metastatic melanoma.

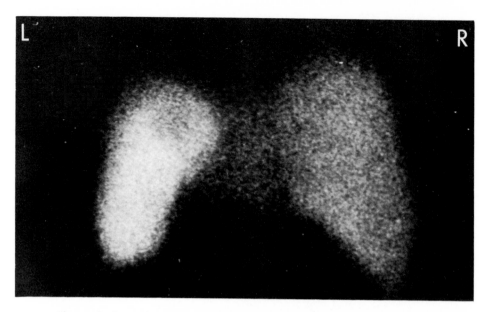

Figure 8. Posterior static image of liver-spleen with Tc-99m sulfur colloid shows a metastatic lesion of melanoma in the upper portion of the spleen.

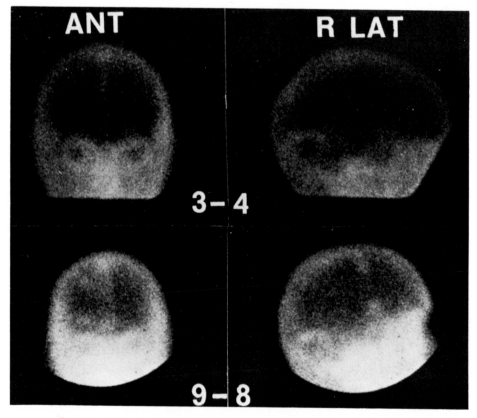

Figure 9. Anterior and right lateral views of 6-month follow-up brain imaging with Tc-99m DTPA show two metastatic lesions of melanoma in the right frontal lobe posteriorly.

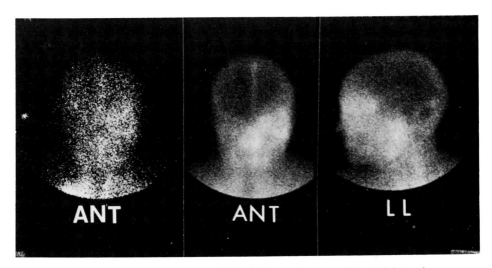

Figure 10. Anterior dynamic study, and anterior as well as left lateral static images of head with Tc-99m DTPA, show increased activities in the left orbit and left frontal lobe inferolaterally in a patient with left lacrimal carcinoma.

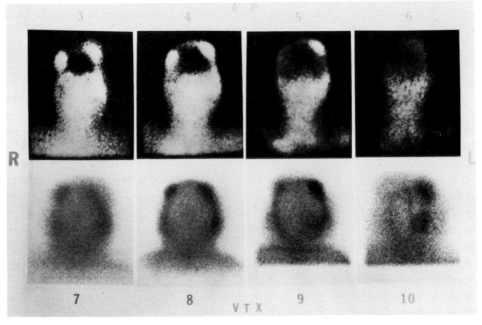

Figure 11. Longitudinal (upper row) and transverse (lower row) sectional images of head with Tc-99m DTPA using Pho/Con camera show metastatic lesions of parotid carcinoma in the scalp.

OTHER RADIOLOGIC PROCEDURES

Plain Films. Routine plain films of the skull, facial bones, neck, and soft tissues of the neck in the frontal and lateral positions are essential in the initial radiologic investigation of head and neck tumors. Even when the plain films do not give a definite diagnosis, they will give a perspective on the clinical problem with respect to the surrounding anatomic structures. They are also important for comparison when interpreting tomograms, laryngograms, or CT scans.

Xeroradiography. Xeroradiography will bring out the soft tissues of the neck in greater detail than the soft tissue x-ray technique. It is especially useful in the demonstration of masses in the nasopharynx, posterior portion of the oropharynx, and hypopharynx. Its use also has been increasing especially in demonstration of the larynx and trachea.

Tomography. Tomography of the skull and facial bones is most useful for defining the extent of primary tumors. Tomography of the larynx is most useful in the investigation of laryngeal tumors when the large exophytic supraglottic type obscures the vocal cord or extends into the subglottis. It is also useful in identifying the site and extent of a tracheal stenosis.

Contrast Laryngotracheography. By laryngography the size of the tumor can be determined for all sites within a mean variation from 0.3 to 0.6 cm depending on the tumor site. Laryngography is also important to assess the fixation of the laryngeal structures for staging.

Computed Tomography. Although sporadic reports have been made about the use of CT in the study of head and neck tumors, there have been no definitive studies in the literature. The probable reason for this is the easy accessibility of these structures by the use of simpler radiologic procedures. The advantages of CT scanning are the ability to obtain axial views to see the relationship of the lesion to surrounding organs and the ability to analyze the tissue morphology and tissue type. The method is also relatively nontoxic and noninvasive. High resolution CT provides excellent imaging for evaluating aggressive lesions of the orbit and paranasal sinus.

Angiography. In suspected cases of vascular tumors or occlusions related to tumors, aortography and selective studies of the common, external, and internal carotid arteries, subclavian, and vertebral arteries can be utilized.

Sialography. Sialography requires relatively gross structural changes for adequate differentiation between neoplasm and inflammation. Tumors in and adjacent to the parotid may displace salivary ducts, and maligna. ~ors within the parotid gland usually cause irregular filling of the ducts.

REFERENCES

1. Bonuiuk M: A crisis in the management of patients with choroidal melanoma. Am J Opthalmol 1979; 87:840–842
2. Belson TP, Lehman RH, Chobanian SL, et al: Bone and liver scans in patients with head and neck carcinoma. Laryngoscope 1980; 90:1291–1296
3. Cutler SJ, Young JL: Third national cancer survey: Incidence data. Natl Cancer Inst Monogr 1975; 41:10–27, 100–135, 388–427
4. Jensen RD, Miller RW: Retinoblastoma: Epidemiologic characteristics. N Engl J Med 1971; 285:307–311
5. Castro EB, Huvos AG, Strong EW, et al: Tumors of the major salivary glands in children. Cancer 1972; 29:312–317
6. Keller AZ: Histology, survivorship and related factors in the epidemiology of eye cancers. Am J Epidemiol 1973; 97:386–393
7. Salmonsen PC: Combined approach to treatment of retinoblastoma. Tex Med 1975; 71:48–52
8. Blanck C: Carcinoma of the parotid gland: Morphology and long term prognosis. Acta Univ Upsaliensis 1974; 195:1–114
9. Lewis JS, Castro EB: Cancer of the nasal cavity and paranasal sinuses. J Laryngol Otol 1972; 86:255–262
10. Creely JJ, Lyons GD, Trail ML: Cancer of the nasopharynx: A review of 114 cases. South Med J 1973; 66:405–409
11. Schall GL, Dichiro G: Clinical usefulness of salivary gland scanning. Semin Nucl Med 1972; 2:270–276
12. Noyek AM, Prtizker KP, Greyson ND, et al: Familial Warthin's tumor. J Otolaryngol 1980; 9:90–96
13. Noyek AM, Greyson ND, Fernandes BJ, et al: Radionuclide salivary scan imaging of a functioning malignant parotid tumor. J Otolaryngol 1982; 11:83–85
14. Noyek AM, Greyson ND, Cooter N, et al: Radionuclide salivary gland imaging of maxillary sinus oncocytoma. J Otolaryngol 1982; 11:17–22
15. Van den Akker HP, Busemann-Sokole, Becker AE: Acinic cell carcinoma of the sublingual gland scintigraphy in preoperative evaluations. Int J Oral Surg 1981; 10:363–366
16. Teates CD, Preston DF, Boyd CM: Gallium-67 citrate imaging in head and neck tumors: Report of cooperative group. J Nucl Med 1980; 21:622–627
17. Poublon RM: Evaluation of head and neck tumors before and after radiotherapy of 40 GY using Ga-67 citrate. A pilot study. Clin Otolaryngol 1982; 7:181–184
18. Senjukov MV, Kulikov VA, Matrejenko EG, et al: Clinical radiological and nuclear medicinal methods for the diagnosis of laryngeal tumors. Radiol Diag 1981; 22:799–805
19. Woolfenden JM, Alberts DS, Hall JN, et al: C0-57 bleomycin for imaging head and neck tumors. Cancer 1979; 43:1652–1657
20. Bergstedt HF, Lind MG, Silfversward C: Facial bone scintigraphy diagnosis of malignant lesions in the mandible. Acta Radiol 1981; 22:485–493
21. Bergstedt HF, Lind MG: Facial bone scintigraphy. Diagnosis of malignant lesions in the maxillary, ethmoidal and palate bones. Acta Radiol 1981; 22:609–618
22. Salvatore M, Avitabile G, Muto V: Scintigraphy in the problem of metastases from cancer of the larynx. Rev Laryngol Otol Rhinol 1981; 102:535–541
23. Shields JA: Accuracy and limitation of the P-32 test in the diagnosis of ocular tumors: An analysis of 500 cases. Ophthalmol 1978; 85:950–966
24. Lanning R, Shields JA: Comparison of radioactive phosphorus (^{32}P) uptake test in comparable sized choroidal melanomas and hemangiomas. Am J Ophthal 1979; :769–772

CHAPTER 4

Lung and Mediastinal Tumors

INTRODUCTION

Cancer of the lung is a lethal disease. When first seen, 50 percent of all patients with bronchial carcinoma are inoperable; of those who are subject to operation, another 50 percent (or 25 percent of the original group) are unresectable. Only 10 percent of patients survive for 5 years. Cure is infrequent because bronchogenic carcinoma metastasizes early. From the time of diagnosis, the average overall survival is 6 to 9 months, with only 20 percent of patients surviving 1 year. Little progress has been made in the past few decades in improving means of early detection or of treatment methods for lung cancer. Progress in other sites has done little to alter the grim survival statistics of this disease. The investigative interest is in early detection, and in combining standard treatments with chemotherapy.

Ga-67 citrate has been found to have a high affinity for lung and mediastinal tumors, and Ga-67 imaging has been helpful in preoperative evaluation of hilar and mediastinal involvement in lung cancer. It has also been useful in early detection of diffuse carcinomatosis and chemotherapy-related interstitial pneumonitis or pneumocytis carinii pneumonia before radiography becomes abnormal. Radionuclide brain, liver, and bone imaging have been utilized for the workup of distant metastases, since those organs are most commonly involved secondarily. Superior vena cava obstruction occurs in 5 percent of lung cancer cases, and is managed as a radiotherapeutic emergency. The Tc-99m venogram has been used to confirm a significant obstruction of the superior vena cava or other abnormal pulmonary venous circulation. Radionuclide studies of pulmonary perfusion and ventilation have been found to objectively assess the resectability and operability for the lung cancers. Radionuclide blood pool imaging of the heart or Tl-201 myocardial imaging occasionally detects myxoma in the atrium.

CHARACTERISTICS OF LUNG AND MEDIASTINAL NEOPLASMS

Epidemiology

Carcinoma of the lung has become the most frequent form of cancer in men in the United States and many other countries. The estimated number of new lung cancer cases in the United States for 1983 is 135,000, constituting 22 percent of new cancer cases in males. The estimated deaths from lung cancer for 1983 is 35 percent of estimated cancer deaths in males.[1] Female lung cancer rates rose rapidly in the 1960s and 1970s, and now constitute 9 percent of new cancers and 17 percent of cancer deaths in women in 1983. This fantastic increase in cancer of the lung has resulted in many extensive clinical, epidemiologic, and laboratory investigations. There appear to be many causative factors related to the genesis of lung cancer, and most of these share a part in the problem. It is obvious, however, that the greatest risk appears to be related to tobacco smoke inhalation,[2] although there is obviously a variation in individual susceptibility to carcinogenesis from this agent.

Primary tumors of the mediastinum are uncommon; about one-third of them are malignant. Teratomas and neurogenic tumors are prevalent in patients 10 to 30 years old, whereas thymomas occur infrequently in young patients.[3] Malignant tumors of the mediastinum predominate in elderly individuals. A variety of rare tumors, including tumors of the heart, make up small proportions of the total of these tumors.

Pathology and Metastasis

An overwhelming proportion of lung tumors arise within the bronchi, but a few malignant neoplasms arise from bronchiolar epithelium and pleural mesothelium.

Bronchial carcinoma is found more frequently in the right (60 percent) than in the left lung, and it arises in a major bronchus in about 75 percent of the cases.[4] A collected series of 2754 carcinomas of the lung were divided as follows: squamous cell carcinomas, 48 percent; undifferentiated carcinomas, 31 percent; adenocarcinoma, 21 percent.[4] Undifferentiated carcinomas included small (oat) cell and large cell types. Bronchioalveolar carcinomas may present as a solitary nodule, or as a multinodular alveolar or reticular manifestation.[5] Giant cell carcinomas of the lung comprise about 6 percent of pulmonary carcinomas. Bronchial adenoma is a misnomer, for these tumors infiltrate locally and metastasize to regional lymph nodes and, occasionally, to distant viscera.[6]

Bronchial carcinoids arise predominantly in the main-stem bronchi with a preponderance for the right lower lobe. Metastasis in the liver may become manifest as a carcinoid syndrome.[7] Sarcomas of the lung are rare. Benign tumors found in the lung include chemodectomas and clear cell tumors that do not metastasize. The profuse pulmonary network of lymphatics, and the great vascularity and constant movements of the lung, facilitate the spread of bronchial carcinomas. The regional lymphatic spread to mediastinal and peritracheal lymph nodes takes place in the majority of cases. The lymphatic spread becomes more extensive when pleural adhesions form and distant pathways of dissemination are facilitated. The tumor may involve the periesophageal, para-aortic, and pararenal nodes through the diaphragm. Spread to the adrenal glands takes place through the lymphatics. Auerbach et al.[8] found that 96 percent of lung cancer patients had metastases at autopsy, most frequently in lymph nodes, but they also found 45 percent with metastasis in the brain, 44 percent in the liver, 34 percent in the adrenal glands, 30 percent in the bones, 24 percent in the kidneys, 13 percent in the pancreas, 10 percent in the spleen, 7 percent in the thyroid gland, and 3 percent in the skin. Spread may take place through the vertebral vein plexus to the scalp. Small cell undifferentiated carcinomas are those with most frequent widespread metastases at autopsy (99 percent).

Teratomas are the most common tumors of the mediastinum. They usually are located in front of the pericardium and great vessels. Most of these tumors are benign, but a few are malignant. Neurogenic tumors are the tumors most frequently found in the posterior portion of the mediastinum. Thymic tumors constitute about 10 percent of mediastinal tumors, and they usually arise in the anterior mediastinum. Lipomas may arise in the anterior or posterior portion of the mediastinum, and pheochromocytomas may arise in the posterior portion of the mediastinum.

Mediastinal germinomas arise in the anterior portion of the mediastinum of young men, usually in the third decade of life. Primary tumors of the heart are rare, and the majority of these are benign. The most common is the myxoma, usually polypoid, which arises frequently in the left atrium. Leiomyosarcomas may arise from the inferior vena cava or peripheral veins.

Malignant teratomas metastasize to mediastinal lymph nodes and to the lungs. Malignant neurogenic tumors may metastasize to the lungs, liver, and other organs after they infiltrate directly neighboring organs. Malignant thymomas rarely metastasize, but metastases involving the lymph nodes, lungs, liver, brain, and kidneys have been reported.[9]

Clinical Evolution

Carcinomas of the lung are among the most insidious of all neoplasms. The earliest symptom is usually an irritative cough, accompanied by mucoid expectoration. Sometimes hemoptysis may occur, and repeated attacks of obstructive pneumonitis may develop over a period of weeks or months. Pleural effusion and empyema may develop with constant pain and dyspnea. Tumors developing in the periphery of the upper lobes may produce an early Horner's syndrome due to compression of the cervical sympathetic plexus. Patients with bronchial carcinoma may present extrapulmonary manifestations not suggestive of the primary cause. These may be metabolic, neuromuscular, vascular, hematologic, osseous, or connective tissue abnormalities.[10] Bronchial carcinoid is so vascular that hemoptysis is often the first symptom.

Mediastinal tumors produce symptoms that are due to pressure and depend on their location. In the anterior portion of the mediastinum, the most common symptoms produced are retrosternal pain, dyspnea, and respiratory complications. In the posterior portion, there may be compression of the trachea or bronchi, with resulting cough and dyspnea, or compression of the phrenic or recurrent nerves with consequent paralysis of the diaphragm or of the larynx. Compression of the esophagus results in dysphagia.

Atrial myxomas may cause pulmonary hypertension and lower cardiac output. Malignant tumors of the heart can also cause congestive heart failure.

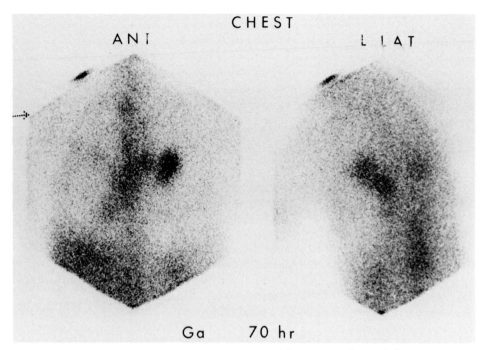

Figure 1. Anterior and left lateral views of the chest 70 hours following the injection of Ga-67 citrate show a squamous cell carcinoma in the left hilum. Also noted are lesions in the right lung.

NUCLEAR IMAGING

Imaging for Primary Tumors

Gallium-67 Imaging. Imaging of the chest with Ga-67 citrate is relatively easier to perform and interpret than imaging of the abdomen, because normally pulmonary concentration is low after 48 hours and physiologic accumulation in bones and breast can be recognized by its distribution. Modern large field cameras with multiple pulse-height analyzers give substantially better gallium images than those available in the past. It has become clear that gallium uptake is nonspecific and occurs in inflammatory lesions as well as neoplasms. Both acute bacterial and chronic granulomatous infections may show uptake. This nonspecificity reduces the utility of gallium imaging in differential diagnoses, but does not detract from its value in localization of focal disease and assessment of its extent, activity, and treatment response.

Ga-67 imaging showed 90 percent positivity in 264 cases with untreated primary lung cancer, and 80 percent positivity in 149 cases with untreated pulmonary lymphoma (Figs. 1 and 2).[11] In lung cancer the histologic type of the lesion appears to have no significant effect on the probability of lesion detection.[12] However, differential uptakes were found for the various tumor types, with anaplastic small-cell carcinoma having the greatest average uptake, and adenocarcinoma anaplastic large-cell carcinoma the smallest.[13] It was suggested that the greater the Ga-67 uptake in the tumor, the more effective is radiation therapy in reducing tumor size. However, the response

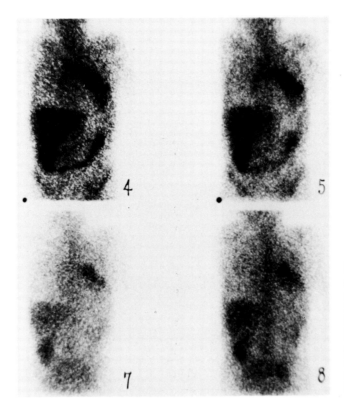

Figure 2. Longitudinal section images of the chest and abdomen with Ga-67 citrate using Pho/Con camera at 48 hours show a squamous cell carcinoma in the left hilum and metastatic lesions in the left lower lung.

to treatment by the quantitative study of Ga-67 uptake did not help in predicting the eventual prognosis.[14]

It has also been reported that Ga-67 imaging gives more consistent and reliable results in cases of squamous cell carcinoma than in cases of adenocarcinoma or alveolar cell carcinoma.[15]

Ga-67 scan has been employed for the evaluation of resectability of lung cancer in patients with collapse and/or consolidation of the involved area as seen on the chest radiograph.[16] Primary mediastinal seminomas[17] and mesotheliomas[18] have also been detected by Ga-67 scanning (Figs. 3 and 4).

Co-57 or In-111 Bleomycin Imaging. In a study of 53 lung cancer patients using 500 uCi Co-57 bleomycin, the tumor-to-lung field radioactivity ratios were greater than 1.5.[19] It was noted that Co-57 bleomycin was a much more effective image-enhancing agent than Ga-67 citrate. However, the true positive rates of Ga-67 citrate, In-111 bleomycin, and Tc-99m citrate for the evaluation of 86 pulmonary malignancies were 79 percent, 65 percent, and 46 percent, respectively.[20] Co-55 bleomycin using a positron camera also detected lung cancers,[21] but the tumor-to-lung activity ratios were lower than those produced by Co-57 bleomycin.

Tc-99m Glucoheptonate Imaging. Tc-99m glucoheptonate was accumulated in 23 of 26 primary lung cancers.[22] Only one of eight benign lung processes

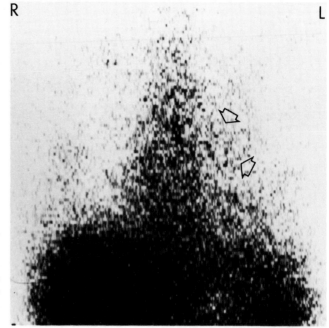

Figure 3. Anterior image of the chest with Ga-67 citrate at 48 hours shows some uptake of radiogallium in the thymoma (arrows), corresponding to the mass lesion seen on chest radiograph.

was visualized, and 23 patients without lung disease had no pathologic foci. The specificity of Tc-99m glucoheptonate scan for tumor detection was higher than that of chest radiographs.

Radionuclide Ventilation-Perfusion Lung Scans. Xe-133 gas ventilation and Tc-99m HAM perfusion studies (Fig. 5) in 45 patients with unresectable cancer of the bronchus were always abnormal in the lung affected by the

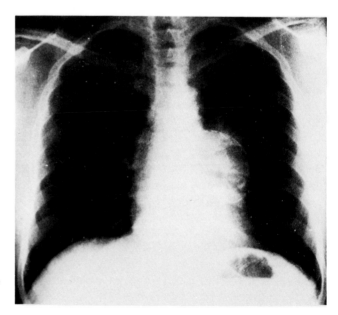

Figure 4. See legend for Fig. 3.

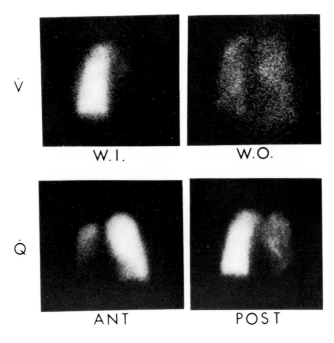

$\dot{V}$

W. I.   W. O.

$\dot{Q}$

ANT   POST

Figure 5. Wash-in and wash-out images of Xe-133 ventilation study (V̇) show severe obstruction of left main bronchus, and generally decreased perfusion (Q̇), especially left lower lobe.

tumor; perfusion was usually more impaired than ventilation.[23] These abnormalities were difficult to detect or to evaluate from the standard chest radiograph. After radiotherapy, ventilation improved in 83 percent and perfusion in 86 percent of the patients. In 87 patients with early lung cancer, 75 percent had a perfusion defect and 67 percent had a ventilation abnormality at the tumor site.[24] Twenty-seven patients with suspected cancer, who were subsequently proved to have benign lesions, also had similar abnormalities, suggesting a limited role of ventilation-perfusion studies in the differential diagnosis of lung cancer.

Tumor emboli showing ventilation-perfusion mismatch occur more frequently than would be suspected clinically.[25]

Radionuclide Angiocardiography. Radionuclide angiocardiography has demonstrated atrial myxomas[26,27] and myocardial rhabdomyomas[28] as filling defects or masses displacing cardiac chambers. Carotid body chemodectomas, which are pulsatile and vascular, have also been diagnosed by dynamic radionuclide scanning.[29]

Imaging for Metastatic Tumors

Ga-67 Imaging. Ga-67 scan (Figs. 6 to 8) has been valuable in separating primary from secondary lung tumors, determining the extent of contralateral hilar or mediastinal lymph node involvement, and detecting distant organ involvement. In a study with 100 lung cancer patients, Ga-67 sensitivity for primary lung cancer was 96 percent, and was 75 percent for detecting distant organ metastases.[30]

In another study of 70 lung cancer patients having Ga-67 positive lung lesions, mediastinal Ga-67 uptake had a sensitivity of 88 percent and a specificity of 86 percent.[31] Can Ga-67 scan be used as a noninvasive tool to

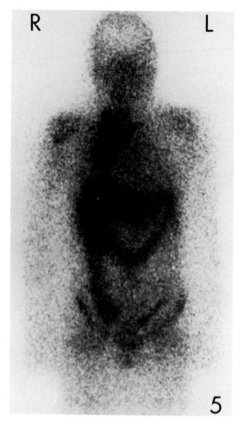

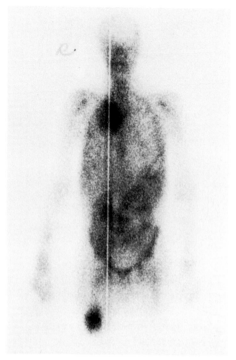

Figure 6. Longitudinal section image of whole body 48 hours following injection of Ga-67 citrate shows a large squamous cell carcinoma in the right hilum, and two metastatic lesions in the right lower neck.

Figure 7. Anterior whole-body image at 48 hours with Ga-67 citrate shows a large adenocarcinoma in the right upper lung, corresponding to a large mass seen on chest radiography. Incidentally noted is a metastatic lesion in the right femur.

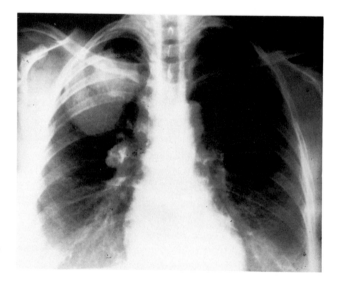

Figure 8. See legend for Fig. 7.

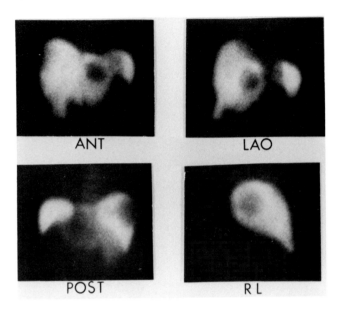

ANT   LAO

POST   R L

Figure 9. Routine static views of liver-spleen with Tc-99m sulfur colloid show metastatic lesions of lung cancer in both right and left lobes.

help determine whether staging mediastinoscopy or thoracotomy should be performed on patients with possible mediastinal extension? If the primary lung cancer is Ga-67 positive, a negative mediastinal scan might obviate against mediastinoscopy. However, it has been reported that although Ga-67 scan is accurate if positive, it is of marginal value if negative.[32] On the other hand, if a mediastinum is Ga-67 positive, mediastinal exploration is still indicated. Gallium-bone (using Tc-99m MDP) subtraction study increases sensitivity for mediastinal disease detection without substantially altering specificity.[33] Ga-67 scan was more accurate than chest radiography in assessing regional nodes (overall accuracy 85.3 percent versus 56 percent).[34] When positive, both procedures correctly indicate malignant involvement of regional nodes.

Ga-67 citrate accumulation in the myocardium and pericardium with histiocytic lymphoma, melanoma, and angiosarcoma has been reported,[35] and a pericardial metastasis of lung cancer has also been demonstrated on Ga-67 scan.[36]

Radionuclide Liver, Bone, and Brain Scans. In carcinoma of the lung, liver and bone scans (Figs. 9 to 12) are frequently indicated. Examples are patients who are either symptomatic or have positive physical findings on laboratory tests. In the asymptomatic patients, some authorities restrict bone and liver scanning in suspected bronchogenic carcinoma to stages II and III, while performing both tests in small cell or adenocarcinoma irrespective of stage.[37] However, other authors have reported as many as 15 percent of asymptomatic patients with positive liver-spleen scans preoperatively, and between 14 and 36 percent with bone metastases.[38] Abnormal bone or liver-spleen scans in the preoperative evaluation of lung cancer patients may prevent unnecessary thoracotomies. The CT head scan is the procedure of choice in patients with neurologic signs and symptoms because its high resolution capabilities provide anatomic detail to an extent not possible with current radionuclide brain imaging. However, radionuclide brain imaging (Fig. 13) has comparable

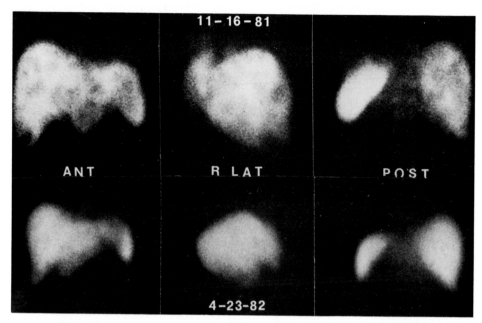

Figure 10. Tc-99m sulfur colloid liver-spleen images show significant improvement of multiple metastases of lung cancer in both lobes of the liver 6 months following chemotherapy.

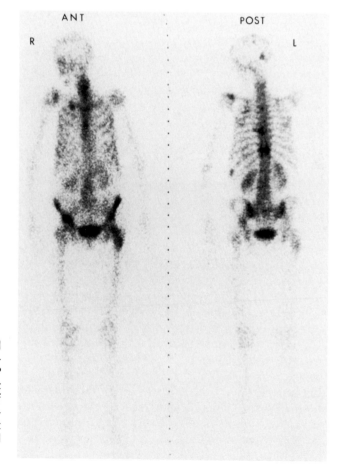

Figure 11. Anterior and posterior whole-body images with Tc-99m MDP show multiple metastatic lesions (focal hot spots) of lung cancer in skull, thoracic vertebrae, ribs, right scapula, and left proximal femur.

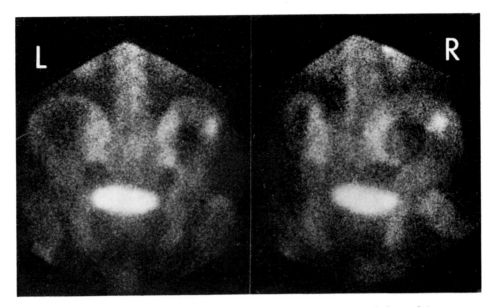

Figure 12. Posterior and right posterior oblique views of the pelvis with Tc-99m MDP show metastatic lesions (focal hot and cold spots) of lung cancer in right iliac bone.

sensitivity to CT with respect to detection of cerebral metastases. In follow-up of known brain metastases, brain scanning may in fact be preferable to CT because of cost considerations (Fig. 14).

In following patients with bronchogenic carcinoma of the lung, liver-spleen and bone scans (Figs. 15 to 17) are usually performed in patients who are either symptomatic or who have abnormal physical findings on laboratory

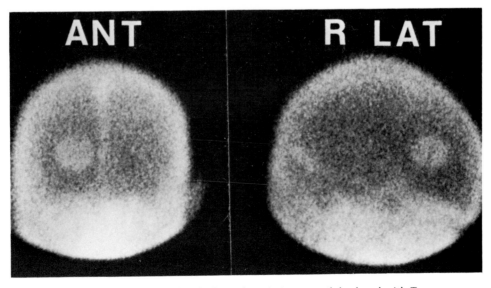

Figure 13. Anterior and right lateral static images of the head with Tc-99m DTPA show a metastatic lesion of lung cancer in the right frontal lobe.

6 / 81

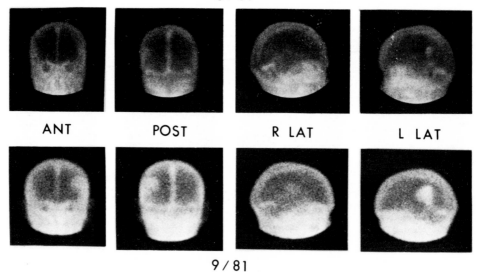

ANT   POST   R LAT   L LAT

9 / 81

Figure 14. Follow-up brain imaging with Tc-99m DTPA shows significant progression of metastatic lung cancer in the left parietal lobe at 3 months.

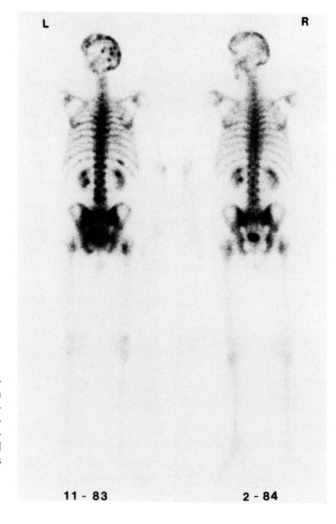

Figure 15. Posterior images of axial skeleton with Tc-99m MDP show significant improvement of metastatic lesions of lung cancer in vertebrae and posterior ribs 2 months following chemotherapy.

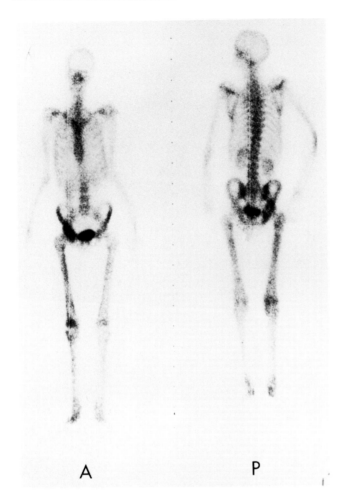

Figure 16. Anterior and posterior whole-body images with Tc-99m MDP show unusual metastatic lesion of lung cancer in the right acetabulum and increased radioactivity along the femoral cortex bilaterally, suggesting pulmonary osteoarthropathy.

tests. In small cell or adenocarcinoma, many protocols call for bone and liver scans every 4 to 6 months because of the high likelihood of relapse.[39] In addition, some authorities perform bone scans in following patients with bronchogenic carcinoma. Once metastatic foci are determined, the bone scan has proven to be useful in assessing the efficiency of therapy.[40] Clinically indicated scans were frequently found to be of more benefit, and negative scans were as useful as positive scans if clinically indicated.[41]

Tc-99m Phosphate Imaging. Bone scans using Tc-99m pyrophosphate revealed abnormal uptake of the radioactivity in primary anaplastic lung cancer and hepatic metastastases.[42] Metastatic lesions within the myocardium were also demonstrated by Tc-99m pyrophosphate imaging.[43]

Tl-201 Imaging. Myocardial scan using Tl-201 (Fig. 18) in a patient with lung cancer demonstrated a sharply bounded defect in the anterolateral and posterior wall of the heart.[44] A metastatic tumor in the pericardial space with extensive myocardial infiltration was found at surgery.

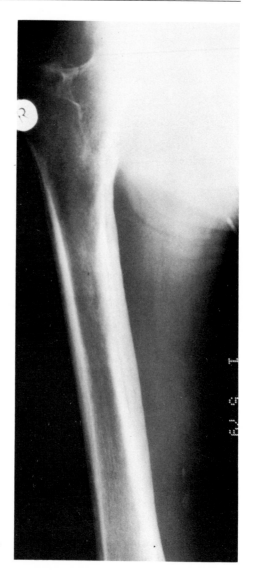

Figure 17. Radiograph of right femur also shows a cortical thickening.

Lymphoscintigraphy. An injection of 30 to 50 uCi Au-198 was given bilaterally via the needle of a flexible bronchofiberoscope into the mucosal or submucosal membranes of the B_8 or B_9 bronchi of 11 patients suspected of lung cancer.[45] In cancer with right hilar or carinal lymph node infiltration, the right mediastinal or carinal lymph nodes were not visualized by scintigraphy. In bronchitis or pneumonia, the carinal or right mediastinal lymph nodes were clearly visualized at 24 to 72 hours.

Other Nuclear Procedures

The Tc-99m venogram (Fig. 19) has been found to be useful for the optimum radiation schedule on the treatment of superior vena cava obstruction.[46] The results of pretreatment Tc-99m venograms correlated with those of serial chest radiographs.

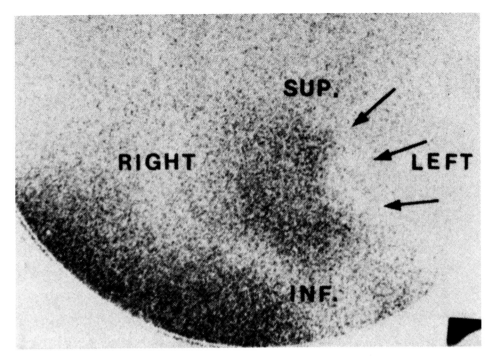

Figure 18. Anterior myocardial image with T1-201 at rest shows a sharply demarcated area of markedly diminished uptake (arrows), corresponding to metastatic tumor infiltration found at surgery. (*From Chest 1980; 78:99, with permission.*)

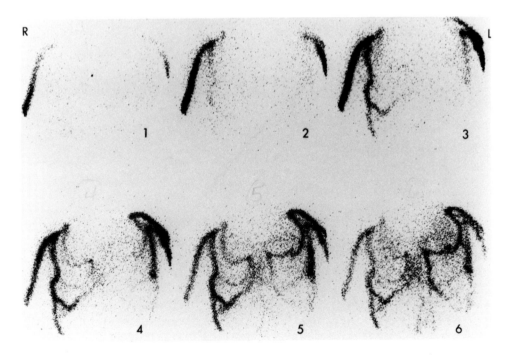

Figure 19. Sequential dynamic images of the chest following the injection of Tc-99m HAM into hand veins bilaterally show complete obstruction of right and left subclavian veins by a large upper mediastinal tumor with multiple collateral circulations along the chest walls.

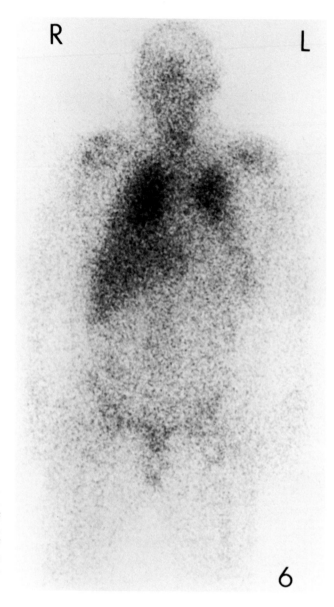

Figure 20. Anterior whole-body image 48 hours following the injection of Ga-67 citrate shows diffuse abnormal uptake of radiogallium in both lungs in a patient with lung cancer and chemotherapy.

Assessment of late left ventricular dysfunction (measurement of ejection fraction) by radionuclide cineangiogram has been useful to detect a doxorubicin cardiotoxicity.[47]

Quantitative ventilation-perfusion lung scans can be useful evaluative procedures. The predicted postoperative forced expiratory volumes (FEV_1) calculated from the results of spirometry with the quantitative measurement of differential perfusion and/or ventilation according to several equations, have been comapred to the observed postoperative FEV_1 in 37 patients, and it was found that FEV_1 after pneumonectomy can be predicted as accurately from the perfusion scan as from the ventilation scan.[48] Patients with a calculated postoperative FEV_1 of less than 1.01 are considered medically inoperable.

Lung images with Ga-67 citrate have revealed acute radiation changes and chemotherapy toxicity.[14] Some cases showed diffuse abnormal radioactivities without corresponding radiographic abnormalities (Fig. 20). A significant reduction in the intensity of uptake can be observed after the chemotherapy or radiation is held. Thus, Ga-67 imaging gives a useful quantitative measure of the response to therapy prior to radiographic and clinical changes.

OTHER RADIOLOGIC IMAGING

Roentgenologic Examination. The chest x-ray is the most important of all methods employed in the diagnosis of lung or mediastinal tumors. Adequate utilization implies not only radiographs in several projections, radiographs in forced expiration, and spot and overexposed x-rays, but also the use of air and opaque substances for contrast and of special radiography, such as stereography, planigraphy, and angiography.

Bronchography has been almost replaced by a wider use of planigraphy (tomography or laminography). Pulmonary angiography is of value in determining major vessel involvement and resectability in carcinomas of the hilum. Fluoroscopy of the heart may be helpful in evaluating patients with tumors of the heart.

Computed Tomography. There is general agreement that CT is more sensitive but less specific than whole lung tomography in detecting pulmonary metastases.[49] That CT does identify more nodules is agreed. This would be expected because of the elimination of confusing shadows caused by superimposed cardiovascular and bony structures present with whole-lung tomography. CT is the method of choice in the evaluation of suspected mediastinal adenopathy.

Echocardiogram. Echocardiography is a well accepted diagnostic tool in the anatomical and functional evaluation of patients with suspected cardiovascular disease. An M-mode echocardiogram detected a cardiac tumor posterior to the tricuspid valve.[50]

REFERENCES

1. Silverberg E: Cancer statistics, 1983. Ca-A Cancer J Clin 1983; 33:9–25
2. Oettlé AG: Cigarette smoking as the major cause of lung cancer. S Afr Med J 1963; 37:957–963
3. Wychulis AR, Payne W Sp, Clagett OT, et al: Surgical treatment of mediastinal tumors. J Thorac Cardiovasc Surg 1971; 62:379–391
4. Fountain CF, Carr DT, Anderson WAD: A system for the clinical staging of lung cancer. Am J Roentgenol 1974; 120:130–138
5. Martinez LO, Cohen GH: Bronchioalveolar carcinoma of the lung. South Med J 1974; 67:447–455
6. Tolis GA, Fry WA, Head L, et al: Bronchial adenomas. Surg Gynecol Obstet 1972; 134:605–610
7. Ricci C, Petrassi N, Massa R, et al: Carcinoid syndrome in bronchial adenoma. Am J Surg 1973; 126:671–677

8. Auerbach O, Garfinkel L, Parks VR: Histologic type of lung cancer in relation to smoking habits, year of diagnosis and sites of metastases. Chest 1975; 67:382–387
9. Trell E, Rausing A: Cardiovascular complications in malignant thymoma. Acta Med Scand 1972; 192:559–564
10. Knowles JH, Smith LH, Jr: Extrapulmonary manifestations of bronchogenic carcinoma. N Engl J Med 1960; 262:505–510
11. Siemsen JK, Grebe SF, Waxman AD: The use of gallium-67 in pulmonary disorders. Semin Nucl Med 1978; 8:235–249
12. Beckerman C, Hoffer PB, Bitran JD, et al: Ga-67 citrate imaging studies of the lung. Semin Nucl Med 1980; 10:286–301
13. Higashi T, Wakao H, Nakamura K, et al: Quantitative Ga-67 scanning for predictive value in primary lung cancer. J Nucl Med 1980; 21:628–632
14. McCready VR, Flower MA, Smythe J: Quantitative studies of gallium-67 uptake in human lung tumors during chemotherapy. J Nucl Med 1982; 23:97–101
15. Alazraki N: Usefulness of gallium imaging in the evaluation of lung cancer. CRC Crit Rev Diag Imaging 1980; 13:249–267
16. Hirleman MT, Yiu-Chiu VS, Chiu LS, et al: The resectability of primary lung cancer: A diagnostic staging review. J Comput Tomo 1980; 4:146–163
17. Polansky SM, Barwick KW, Ravin CE: Primary mediastinal seminoma. Am J Roentgenol 1979; 132:17–21
18. Wolk RB: Ga-67 scanning in the evaluation of mesothelioma. J Nucl Med 1978; 19:808–809
19. Takimoto M, Yokoi H, Sakurai T, et al: Evaluation of Bayer map on diagnosing of Co-57 bleomycin scintigram of the lung. Radioisotopes 1979; 28:772–774
20. Apostolidis P, Kostaki P, Hatzigeorgiou N, et al: Comparison of Ga-67 citrate, In-111 bleomycin and Tc-99m citrate in the evaluation of pulmonary lesions. Eur J Nucl Med 1980; 5:57–61
21. Nieweg OE, Bekkhuis H, Paans AM, et al: Detection of lung cancer with Co-55 bleomycin using a positron camera. A comparison with Co-57 bleomycin and Co-55 bleomycin single photon scintigraphy. Eur J Nucl Med 1982; 7:104–107
22. Vorne M, Sakki S, Jarvi K, et al: Tc-99m glucoheptonate in detection of lung tumors. J Nucl Med 1982; 23:250–254
23. Frazio F, Pratt TA, McKenzie CG, et al: Improvement in regional ventilation-perfusion after radiotherapy for unresectable cancer of the bronchus. Am J Roentgenol 1979; 133:191–200
24. Katz RD, Alderson PO, Tockman MS, et al: Ventilation-perfusion lung scanning in patients detected by a screening program for early lung cancer. Radiol 1981; 141:171–178
25. Boudreau RJ, Lisbona R, Sheldon H: Ventilation-perfusion mismatch in tumor embolism. Clin Nucl Med 1982; 7:320–322
26. Ducassou D, Brendel AJ, Roudant R, et al: The use of radionuclide methods to detect right atrial myxomas. J Biophyss Med Nucl 1980; 4:309–312
27. Barsotti A, Mariotti R, Bencivelli W, et al: Cardiac tumors: Validity and limitation of non-invasive detection by first-pass and equilibrium radionuclide imaging. Eur J Nucl Med 1981; 6:6–7
28. Starshak RJ, Sty JR: Radionuclide angiocardiography: Use in the detection of myocardial rhabdomyoma. Clin Nucl Med 1978; 3:106–107
29. Peters JL, Ward MW, Fisher C: Diagnosis of a carotid body chemodectoma with dynamic radionuclide perfusion scanning. Am J Surg 1979; 137:661–664
30. DeMeester TR, Golomb HM, Kirchner P, et al: The role of Ga-67 citrate scanning in the clinical staging and preoperative evaluation of patients with cancer of the lung. Ann Thorac Surg 1979; 28:451–464
31. Fosburg RG, Hopkins GB, Kan MK: Evaluation of the mediastinum by Ga-67 scintigraphy in lung cancer. J Thorac Cardiovasc Surg 1979; 77:76–82

32. Richardson JV, Zenk BA, Rossi NP: Preoperative noninvasive mediastinal staging in bronchogenic cancer. Surgery 1980; 88:382–385

33. Waxman A, Romanna L, Berman D, et al: A method for increasing the sensitivity of gallium scintigraphy in the detection of mediastinal metastasis in patients with primary bronchogenic carcinoma. J Nucl Med 1982; 25:98–101

34. Lunia SL, Ruckdeschel JC, McKneally MF, et al: Noninvasive evaluation of mediastinal metastasis in bronchogenic carcinoma: A prospective comparison of chest radiography and gallium-67 scanning. Cancer 1981; 47:672–679

35. Yeh SD, Benua RS: GA-67 citrate accumulation in the heart with tumor involvement. Clin Nucl Med 1978; 3:103–105

36. Joo KG, Carter JE, Seo S: Pericardial metastasis on gallium scan. Clin Nucl Med 1980; 5:37

37. Gutierrez A, Vincent RG, Bakshis and Pakita H: Radioisotope scans in the evaluation of metastatic bronchogenic carcinoma. J Thorac Cardiovasc Surg 1975; 69:934–941

38. Kelly RJ, Cowan RJ, Ferree DB, et al: Efficacy of radionuclide scanning in patients with lung cancer. JAMA 1979; 242:2855–2857

39. Levenson RM, Sauerbaum BJ, Ihde DC, et al: Small cell lung cancer: Radionuclide bone scans for assessment of tumor extent and response. Am J Roentgenol 1981; 137:31–35

40. Donato AT, Ammerman EG, Sullesta O: Bone scanning in the evaluation of patients with lung cancer. Ann Thorac Surg 1979; 27:300–304

41. White DM, McMahon LJ, Denny WF: Usefulness outcome in evaluating the utility of nuclear scans of the bone, brain and liver in bronchogenic carcinoma patients. Am J Med Sci 1982; 283:114–118

42. Que L, Wiseman J, Hales IB: Small cell carcinoma of the lung: Primary site and hepatic metastases both detected on Tc-99m pyrophosphate bone scan. Clin Nucl Med 1980; 5:260–262

43. Quaife MA, Boschult P, Baltaxe HA, et al: Myocardial accumulation of labeled phosphate in malignant pericardial effusion. J Nucl Med 1979; 20:392–396

44. Lubell DL, Goldfarb R: Metastatic cardiac tumor demonstrated by T1-201 scan. Chest 1980; 78:98–99

45. Matsuo M, Ushio K, Nishiyama S, et al: A study of pulmonary hilar and mediastinal lymphoscintigraphy. Radioisotopes 1979; 28:562–567

46. Scarantino C, Salazar OM, Rubin P, et al: The optimum radiation schedule in treatment of superior vena cava obstruction: Importance of Tc-99m scintiangiograms. Int J Rad Oncol Biol Phys 1979; 5:1987–1995

47. Gottdiener JS, Mathisen DJ, Berer JS, et al: Doxorubicin cardiotoxicity: Assessment of late left ventricular dysfunction by radionuclide cineangiography. Ann Int Med 1981; 94:430–435

48. Wernly JA, De Meester TR, Kirchner RT, et al: Clinical value of quantitative ventilation-perfusion lung scans in the surgical management of bronchogenic carcinoma. J Thoracic Cardiovasc Surg 1980; 80:525–543

49. Libshitz HI, North LB: Pulmonary metastases. Radiol Clin North Am 1982; 20:437–451

50. Caralis DG, Kennedy HL, Bailey I, et al: Primary right cardiac tumor. Detection by echocardiographic and radioisotopic studies. Chest 1980; 77:100–102

CHAPTER 5

Breast Cancer

INTRODUCTION

Carcinoma of the breast is the most common cancer in women in the United States. Despite the frequency of this disease, reports in the literature have not been sufficiently definitive to settle the many arguments related to proper therapy, possibly due to the variable nature of breast cancer in each patient. Since more than half of all patients present with or eventually develop metastases, local measures like surgery and radiotherapy, however aggressive, are not the whole answer. Better survival rates will depend on earlier diagnosis and perhaps on new therapeutic techniques such as chemicals or immunologicals, and ultimately will depend on prevention of the disease when pathogenetic mechanisms of carcinogenesis are finally understood. The use of adjunctive chemotherapy in combination with surgery for operable cases has been established.

Physical examination does not replace mammography, and vice versa. The two modalities are able to uncover early carcinoma independently and their combination can be used as screening/detecting methods. In the United States, in the non-screen setting, 50 to 55 percent of breast cancers detected are stage II, and about 3 percent of all cancers are ductal or lobular in situ.[1] The recent staging data indicate that the percent of clinical stage I cancers at detection for those discovered accidentally, by frequent breast self-examination, or by periodic clinical examination, was 27 percent, 38 percent, and 54 percent, respectively.[2]

Pretreatment bone scans are routinely recommended in the current literature and by physician authorities in suspected carcinoma of the breast because of the strong predilection of the disease to metastasize to bone, and the high sensitivity of radionuclide bone imaging in detecting metastatic foci. Many of the major advantages of using the bone scan involve its use in patients with stage I or II breast cancer.[3] It has been established that the presence of bone pain in breast cancer is an unreliable indicator of metastatic disease, especially when patients are receiving adjuvant chemotherapy.[4] It is generally accepted that all patients should have bone scans every 6 to 12 months for at least the first 3 years following surgery, and less often thereafter. Since recurrence of breast cancer is common after 5 years, individual circumstances should determine the importance of follow-up bone scans. It has been suggested that serial quantitative bone scans be done, in preference to radiographs, to assess the response of bone metastases to systemic therapy.[5]

Liver-spleen scanning in the asymptomatic pretreatment breast carcinoma patient is indicated if liver function tests or other parameters suggesting distance spread are abnormal. The incidence of liver metastases is lower than bone metastases and usually the liver metastases appear chronologically later in patients with progressive disease. A baseline liver scan may be useful in the immediate post-treatment period in stage III or IV disease, but in general should be limited to circumstances where there is hepatomegaly or abnormal liver function studies.

Radionuclide brain scanning is of value in documenting the presence of cerebral metastases when clinically suspected. Lymphoscintigraphy of the internal mammary lymph nodes is rapidly becoming an important adjunct to radiation therapy in the delineation of the therapy field.

CHARACTERISTICS OF BREAST CANCER

Epidemiology
Breast cancer has been the leading cause of cancer among females in the United States. It was estimated in 1982 that 26 percent of cancer in women was primarily in the breast and accounted for 19 percent of cancer deaths in women.[6] The 5-year survival rate in the younger age group (15 to 34 years old) was only 53 percent. The most common treatment modalities employed included surgery alone (48 percent or surgery combined with radiation therapy (31 percent).[6]

Pathology and Metastasis
There are many different histologic types of breast cancer which may offer a basis for the choice of treatment and for the prognosis. The following classification* is modified from that proposed by McDivitt, et al.:

---

*Modified from McDivitt RW, et al: Tumors of the breast, in Atlas of Tumor Pathology, 2nd series, Washington, D.C., A.F.I.P., 1967.

a. Carcinoma of mammary ducts
  Papillary carcinoma
  Intraductal carcinoma
b. Infiltrating carcinoma
  Papillary
  Ductal
  Comedo
  Mucinous
  Medullary
  Tubular
  Adenoid cystic
  Metastatic
  Squamous cell
  Apocrine
  Giant cell
c. Carcinoma of mammary lobules
  Lobular carcinoma in-situ
  Infiltrating
d. Paget's Disease
e. Sarcomas
  Cystosarcoma phylloides
  Stromal sarcoma
  Fibrosarcoma
  Liposarcoma
  Angiosarcoma
  Malignant lymphoma
f. Metastases to breast

The growth rate of breast tumors may be extremely variable, and the presence of regional and distant metastases may occur at any time in the patient with breast carcinoma. Carcinoma may reach the axillary nodes through the lymphatics or the axillary nodes may be bypassed, and systemic vascular invasion may occur. The paraclavicular areas are commonly involved and extension to the nodes of the internal mammary chain is frequent in tumors located in the inner quadrants.

An autopsy study[7] of 647 patients who died of breast cancer revealed 70 percent skeletal, 66 percent pulmonary, and 61 percent hepatic metastases. Less commonly seen were metastases to the CNS (25 percent); to the kidneys (13 percent); and to the thyroid gland (20 percent). This study revealed that the chance of metastasis in these less frequently affected organs was increased greatly when the lungs, liver, and bones were involved. In another autopsy series,[8] metastases to the esophagus, stomach, and intestines were present in 16 percent of the cases. The tumor may reach the bones through the vertebral vein system[9] and thus bone metastases may be present without pulmonary metastases. Other patterns of metastatic disease may produce liver, brain, ovary, kidney, and adrenal metastases with few bone metastases.

Cystosarcomas of the breast rarely metastasize to the lymph nodes, but pulmonary metastases occur more frequently. Fibrosarcoma metastasize more frequently to the lungs, liver, and brain.

Clinical Evolution

The most important single presenting sign of breast cancer is a lump that is usually painless. Occasionally, the first symptom is sensation of heaviness in the breast. If metastatic disease precedes the diagnosis of the primary tumor, there may be back pain due to metastases to the vertebrae. Cough, pleurisy, and dyspnea may be symptoms of hydrothorax due to pulmonary and mediastinal metastases. Cardiorespiratory failure is the most common cause of death from breast cancer. Neurologic signs may develop before death, suggesting brain metastases.

Lemon et al.[10] noted a reduced estriol excretion in patients with breast cancer, and extensive studies of urinary estrogen profiles in relation to the risk of breast cancer have been recommended.[11] The binding of estrogens and progesterone to receptor proteins has been implicated in the patients' ability to respond to ablative hormone treatment, surgery, or chemotherapy.

Clinical staging of carcinoma of the breast proposed by Haagenson et al. is of relatively little value due to limited clinical appraisal of important factors, considerable variation from one examination to another, and varied pathologic entities. Nevertheless, a clinical staging may apply to most cases and permit grouping while also allowing for frequent exceptions.

For the purpose of planning follow-up diagnostic procedures, the anatomic region of statistical first recurrence is of utmost importance. A study of 305 post-radical mastectomy patients, with and without postoperative chemotherapy, demonstrated the following distribution of first recurrence sites: integument (31 percent); skeletal (28 percent); respiratory (19 percent); lymphatic (13 percent); digestive (9 percent); ovaries (1 percent); and brain (1 percent).[12]

NUCLEAR IMAGING

The development of a noninvasive technique for the early detection of breast cancer or differentiation of benign and malignant breast lesions has been a long-standing goal of nuclear medicine research. Early attempts at breast tumor detection using Hg-197 were rather unrewarding. In 1966 Whitley et al.[13] first reported the uptake of Tc-99m sodium pertechnetate in breast carcinoma. Cancroft and Goldsmith[14] used a gamma camera to demonstrate focal accumulation of Tc-99m pertechnetate in patients with breast carcinoma. There was no abnormal radioactivity in two patients with benign breast disease. Villarreal et al.[15] subsequently evaluated six patients with breast carcinoma and demonstrated tumor localization of Tc-99m pertechnetate in five. In 30 cases with benign breast disease, however, 6 showed "false positive" results that limit the use of pertechnetate breast scintigraphy for differential diagnosis and for screening purposes. Sixteen cases of breast carcinoma have also been evaluated by Richman, with 14 demonstrating abnormal Tc-99m accumulation.[16] Discrete focal uptake of the radioactivity was the most common finding (Figs. 1 and 2).

Overall, scintigraphy in 86 patients was slightly more accurate than mammography.[17] Considering only those cases where mammography was equivocal, scintigraphy correctly predicted the outcome in 88 percent of cases. Disregarding equivocal studies, scintigraphy was 92 percent sensitive and 91 percent specific for malignant breast disease. These findings suggested that radionuclide breast imaging could be employed in situations where mammography is inconclusive or disagrees with the clinical impression, and to confirm x-ray results that imply biopsy may be avoided. Excellent anatomical correlation with clinical radiographic and histologic localization of breast malignancies may prove useful in the preoperative documentation of the extent of disease or in the evaluation of the response to therapy.

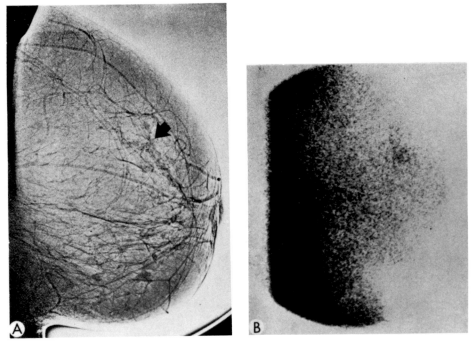

Figure 1. Calcifications (arrow) on x-ray mammogram (**A**) and Tc-99m pertechnetate uptake (**B**) in the breast cancer. (*From Semin Nucl Med 1981; 11:297, with permission.*)

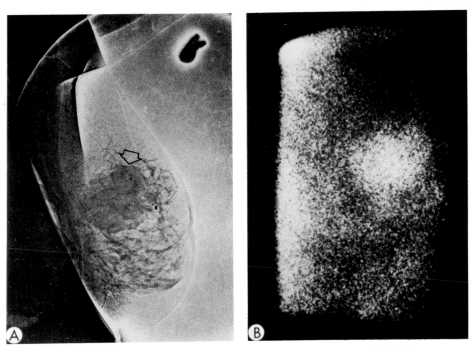

Figure 2. A large mass (arrow) on x-ray mammogram (**A**) and Tc-99m pertechnetate uptake (**B**) in the malignant cystosarcoma of breast. (*From Semin Nucl Med 1981; 11:294, with permission.*)

Gallium-67 has proved a relatively insensitive agent for the detection of primary breast carcinoma. In Richman's experience with ten patients with breast carcinoma, only five had positive Ga-67 scans.[16] When Tc-99m pertechnetate as well as Ga-67 scans were positive, lesions were better visualized with Tc-99m pertechnetate. Ga-67 demonstrated one false-positive in ten cases of benign breast disease.[17] Higasi et al.[18] studied 16 cases of breast carcinoma with Ga-67 and eight were positive. A large study of Ga-67 uptake in the breast in 600 women revealed 18 cases of abnormal uptake of which only four had malignant breast tumors.[19] Normal diffuse uptake in women on cyclic hormones might mask a focal pathologic accumulation of radiogallium.[20]

Limitations of specificity similar to that encountered with Tc-99m pertechnetate have been reported with bone scanning agents. Serafin et al.[21] demonstrated focal accumulation of Tc-99m diphosphonate or polyphosphonate in 82 percent of 17 patients with breast cancer. However, 36 percent of patients with benign breast disease also had positive scans. There was a significant decrease in total and passive uptake of rubidium-86 by red blood cells from 30 breast cancer patients, when compared to 22 controls and eight patients with benign lesions.[22]

Soimakallio and Kiuru[23] made correct scintigraphic diagnosis using ^{111}In bleomycin in 7 of 15 patients with malignant breast tumors, and in 21 of 25 patients with benign tumors. In 32 patients with benign and malignant breast tumors, scintigraphy with Tc-99m bleomycin also showed significantly increased radioactivity in all of the malignant tumors and no accumulation of the radioactivity in the benign tumors.[24] Katzenellenbogen et al.[25] prepared three series of halogenated estrogen analogs as potential breast tumor imaging agents, but no clinical results are yet available.

In summary, breast scintigraphy, although introduced many years ago, has not achieved widespread acceptance, and optimal breast scintigraphy for tumor detection needs more specific radiopharmaceuticals than those presently available. Very recently, in vivo clinical evaluation of human mammary tissue has been reported using the nuclear magnetic resonance (NMR) imaging.[26] Spin-lattice relaxation time (T_1) values of malignant tissue were elevated, and NMR images exhibited lower proton density for mammary dysplasia than normal tissue.

IMAGING FOR METASTASES IN BREAST CANCER PATIENTS

Bone Scan

Since metastatic bone disease is a common cause of recurrence and death in breast cancer patients, the bone scan is frequently employed in the study of patients with breast cancer. Major clinical indications for bone scanning in breast cancer patients are pretreatment staging, as well as follow-up after primary treatment, evaluation of clinically suspicious bony metastases, and monitoring the effects of treatment. There is no doubt that the bone scan is a more sensitive indicator of the presence of bone metastases than the radiograph. In patients with early bone involvement, radiographs are rarely abnormal, while a significant number of cases of occult bony metastases can be detected by scanning (Figs. 3 to 6). In patients with advanced disease, the

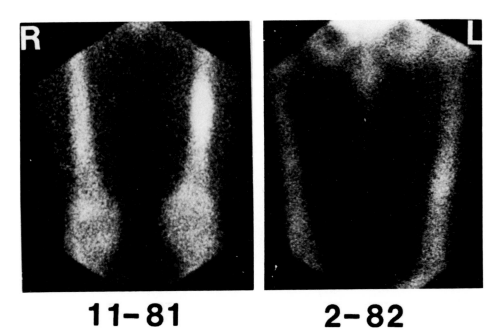

11-81　　　　**2-82**

Figure 3. Anterior images of right and left femurs with Tc-99m MDP show a metastatic lesion of breast cancer in the left femoral shaft. Original radiograph was negative.

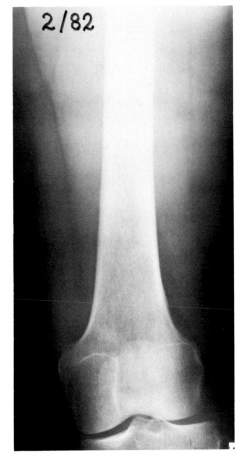

Figure 4. Same femur as in Figure 3. Follow-up bone scan demonstrates some improvement in 3 months following chemotherapy, but radiograph reveals only questionable lytic lesion.

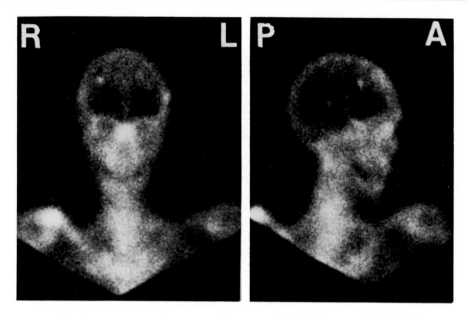

Figure 5. Anterior and right lateral views of skull with Tc-99m MDP show metastatic lesions of breast cancer in right and left frontal skull.

bone scans also have shown a significantly increased number of lesions, compared with the radiograph (Figs. 7 and 8).

However, controversy still exists in the literature concerning the relative value of routine preoperative bone scanning and the metastatic x-ray survey probably based on low occurrence of osseous metastases.[27] McNeil et al.[3]

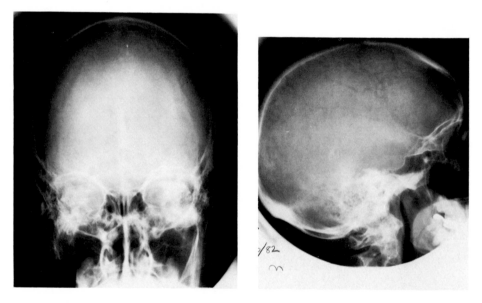

Figure 6. Same skull as in Fig. 5. Radiographs of skull show no definite metastatic lesion.

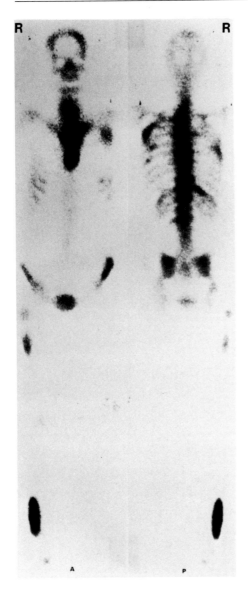

Figure 7. Anterior and posterior whole-body images with Tc-99m HEDP show multiple metastatic lesions of breast cancer in vertebrae, sternum, ribs, left scapula, left iliac crest, right femur, and right fibula. Left lateral chest x-ray shows no definite lesion in the sternum and thoracic vertebrae.

reported that 7 percent of patients with stage I disease and 25 percent of patients with stage II disease develop bony metastases. Half of the patients developed bone metastases within 12 months, and three-quarters of the patients developed bone metastases within 18 months of their initial presentation. Furthermore, a large number of patients (30 to 60 percent) who develop bony metastases as depicted by bone scan (Fig. 9) do not have bone pain.[4,28] Therefore, the bone scan is frequently the first indication of the need to reorient the therapeutic approach.

With the Tc-99m phosphate compounds and high resolution gamma camera, more subtle changes in the bony structure are seen, and the false positive scan is a common problem such as in the case of a solitary rib lesion (Figs. 10 and 11). If abnormal bone scans are correlated with radiographs, and

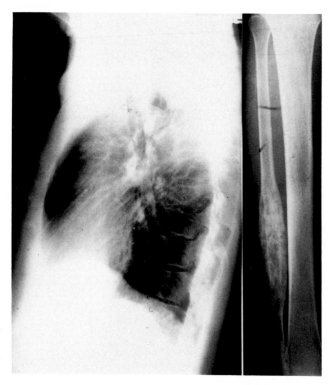

Figure 8. Radiograph of right lower leg (from Fig. 7) shows a lesion in the fibula.

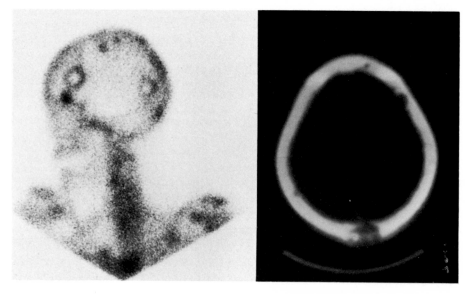

Figure 9. Left lateral view of the head with Tc-99m MDP shows multiple skull metastases of breast cancer in a patient without neurologic symptoms. CT also shows metastatic lesions.

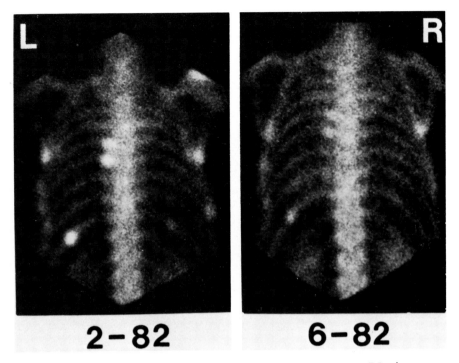

Figure 10. Posterior views of the chest with Tc-99m MDP show multiple rib lesions which appear significantly improved on follow-up scan.

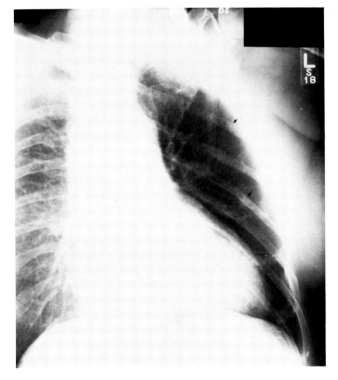

Figure 11. Radiograph of left ribs (Fig. 10) demonstrate healing fractures (arrows).

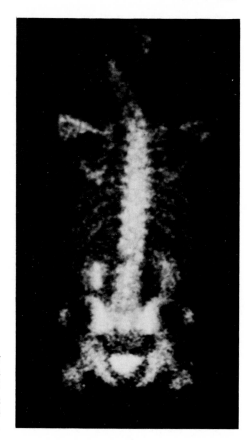

Figure 12. Posterior image of whole-body rectilinear scan with Tc-99m MDP shows a cold metastatic lesion of undifferentiated breast carcinoma in the right aspect of L_2 vertebral body. Also noted are hot lesions in the right eighth rib and occipital skull.

if the scan is interpreted with all the available clinical data, the incidence of true "false-positive" diagnosis should be less than 3 to 5 percent.[28] The bone scan may be clinically useful in cases of clinically suspected bone involvement in helping to localize lesions—e.g., in identifying a suitable site for biopsy, or in designing a radiation therapy field.

Radiation treatment of bone metastases results in reduction of radionuclide uptake and normalization of the scan in the irradiated area. Highly destructive metastases from anaplastic or undifferentiated carcinoma may cause photon-deficient (cold) lesions on bone scan (Fig. 12). Bone scans have a place in the assessment of response to systemic chemotherapy and hormone therapy in patients with disseminated breast cancer (Fig. 13). When bone metastases heal under the influence of treatment, they eventually lose their avidity for radiopharmaceuticals with normalization of the bone scan.

Regression in metastatic lesions as seen on a bone scan (Figs. 14 and 15) may be accompanied by increasing sclerosis radiographically due to progressive osteoblastic tumor deposition or the healing process itself. The presence of active neoplasia can better be determined in sclerotic lesions by the bone scan that demonstrates the tumor-induced new bone formation, as compared to radiographs that show the more unpredictable results of new bone formation, plus the reparative process. A bone scan is therefore useful in

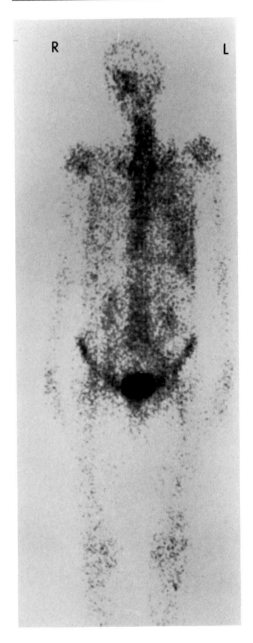

Figure 13. Anterior whole-body bone scan with Tc-99m HEDP shows diffuse radioactivity in the malignant pleural effusion of a patient with right breast cancer.

patients with sclerotic metastases on radiographs to determine whether the disease is active or stable.[5,28,29]

In summary, radionuclide bone scan is now an established, highly sensitive technique in the diagnosis of bone metastases and monitoring the therapeutic effects.

Radionuclide Liver-Spleen Scan

Radionuclide liver scan (Fig. 16) has been the preferred initial screening examination for detection of metastatic disease with diagnostic sensitivity of

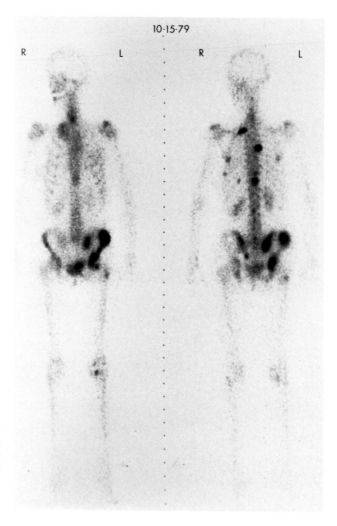

10-15-79

Figure 14. Anterior and posterior whole-body images with Tc-99m MDP in 1979 show multiple metastatic lesions of breast cancer in posterior ribs, vertebrae, pelvic bone, and left proximal femur.

60 to 98 percent and specificity of 64 to 99 percent. Its utility in breast cancer patients is apparent, since metastatic liver disease is present in 61 percent of patients at autopsy.[7] Moreover, 15 to 20 percent of patients die as a result of hepatic decompensation and up to 50 percent have clinically significant metastatic involvement of the liver.[30,31] Current inherent scanning limitations in the detection of metastatic disease require that the lesion be at least 1 to 2 cm in size for visualization.[32]

It is necessary to judiciously interpret the liver scan with all available parameters before diagnosing metastatic liver disease. Correlation of suspicious scans with change in hepatic size and incorporation of liver function tests such as alkaline phosphatase, will enhance the diagnostic accuracy, although it is well known that metastatic disease may be unexpectedly detected in the absence of physical or biochemical abnormalities.[33]

Emission computerized tomographic images have shown more accurate diagnosis of focal defects in the liver in several cases.[34] CT or ultrasound

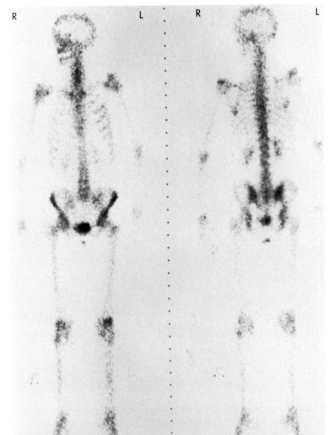

Figure 15. Follow-up scan (from Fig. 7) in 1982 shows a complete resolution of metastases.

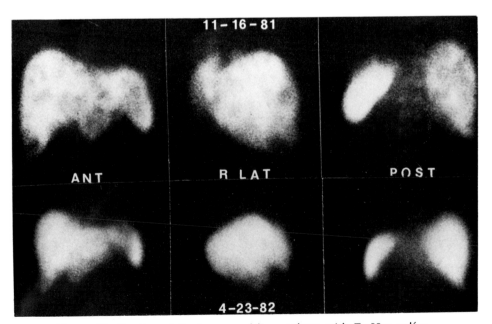

Figure 16. Routine static images of liver-spleen with Tc-99m sulfur colloid demonstrate multiple metastatic lesions of breast cancer in right and left lobes of the liver, and their resolution with chemotherapy on follow-up study.

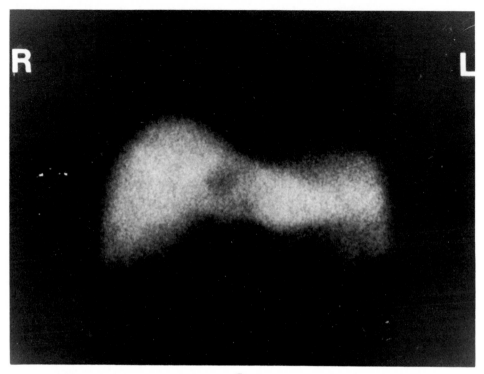

Figure 17. Anterior liver-spleen image with Tc-99m sulfur colloid shows a single defect in the region of porta hepatis in a patient with breast cancer and normal liver function test. Ultrasound confirmed a cyst.

examinations appear to be complementary and helpful when radionuclide studies (Fig. 17) are equivocal in patients with a high clinical suspicion of metastases. At the time of first presentation in 282 patients with breast cancer, the incidence of liver metastases was 13.2 and 23.1 percent by radionuclide scan and ultrasonography, respectively.[35]

The abnormalities detected in liver scintiscans are generally multiple, small, well-defined focal defects, usually starting in the right lobe of the liver, due to its vascularity and size (Fig. 18).

Liver scintiscan has also been a useful adjunct in elucidating the progression of documented metastases and evaluating the response to chemotherapy.

The spleen is routinely visualized on a Tc-99m sulfur colloid liver scan, but no correlation has been observed between spleen scans and the reported 12 percent incidence of splenic metastases at autopsy. It has been suggested that, as in animals, a splenic component of the host reaction to breast cancer responsible for splenomegaly exists in man.[36] In conclusion, Tc-99m sulfur colloid liver scan is recommended for the documentation of suspicious metastasis, assessing tumor progression, and evaluating response to chemotherapy.

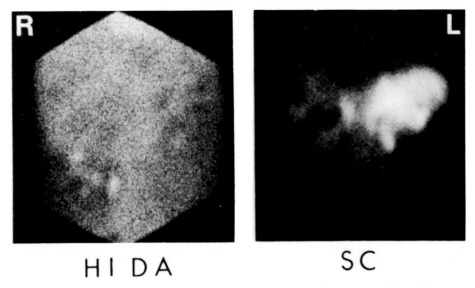

Figure 18. Anterior images of upper abdomen with Tc-99m HIDA and sulfur colloid demonstrate multiple metastatic lesions of breast cancer in both lobes of liver. HIDA scan suggests poor hepatocellular function.

Radionuclide Brain Scan

Brain scan has been properly applied to the breast cancer patient who has neurologic signs or symptoms. The autopsy incidence of breast carcinoma metastatic to brain was 25 percent, while the involvement of the skeleton was 70 percent.[7] The high incidence of skeletal metastases is reflected by the frequent involvement of the calvarium, which can result in an abnormal brain scan. While Tc-99m DTPA or pertechnetate has a better tumor-to-brain ratio than Tc-99m phosphate bone scan agent, the latter is far superior in demonstrating metastases to the bony calvarium. Thus, the use of two scanning agents offers a differentiation of intracerebral from skull metastases.

In the evaluation of patients with suspected intracranial mass lesions, CT is now the preferable initial diagnostic test. However, radionuclide brain scan (Fig. 19) may still serve a satisfactory screening or follow-up examination in certain well-defined clinical situations.

Lymphoscintigraphy

It has been appreciated that local therapy is inadequate in more than 50 percent of patients with resectable carcinoma and histologically involved axillary nodes. The majority of patients succumb to systemic metastases without evidence of local recurrence, suggesting that spread had probably occurred prior to primary surgery. With axillary metastases present, the incidence of parasternal involvement rose to 27.7 percent for tumors of the external quadrants and to 40.5 percent for the inner quadrants.[37] Parasternal lymphatics may constitute a more direct pathway of spread of disease to both

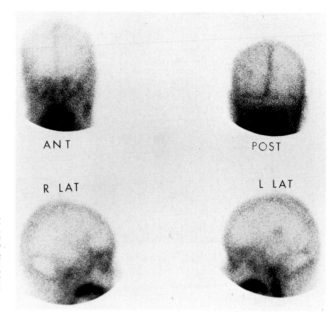

ANT   POST

R LAT   L LAT

Figure 19. Routine static views of brain imaging with Tc-99m DPTA show metastatic lesions of breast cancer in right frontal and left temporal lobes.

mediastinal lymphatics and systemic circulation through communication of the jugulosubclavian junction.

Ege et al.[37] has produced internal mammary lymphangiography data in 1072 patients with breast cancer and reported abnormal studies in 35 percent of patients with, and in 18 percent of patients without, axillary involvement. Unlike surgical extirpation, the technique is simple, can be repeated, and provides a significant noninvasive means of individual patient assessment (Fig. 20).

Deland et al.[38] also reported specific concentration of I-131-labeled antibodies to carcinoembryonic antigen in axillary lymph node metastases from breast carcinoma in seven patients.

In summary, abnormal lymphangiography appears to be helpful in clinical staging of breast carcinoma and the more rational management of this disease.

OTHER RADIOLOGIC PROCEDURES USED IN BREAST CANCER DETECTION

Conventional film mammography[39] and xeroradiography[40] have been the most widely used studies in the evaluation of suspicious breast masses. The indications for mammography are quite broad with the aim being to differentiate a benign breast lesion from malignant disease.

Thermography is another technique for breast examination. The increased vascular and metabolic rate of tumors cause increased heat production resulting in an abnormal thermal pattern. When comparing clinical examination, mammography, and thermography in 306 histologically confirmed breast cancers, the detection sensitivity of these modalities was 82, 85,

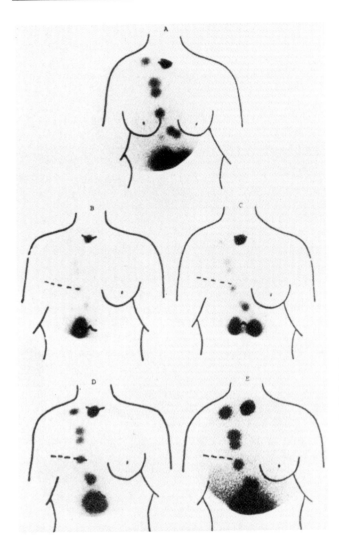

Figure 20. Internal mammary lymphoscintigraphy with Tc-99m antimony colloid. (**A**) Unilateral chain of normal nodes before right mastectomy. (**B, C**) Diminished uptake 1 month after surgery. (**D, E**) Return of normal RES function 5 months after surgery. (*From Int J Radiat Oncol Biol Phys 1977; 2:758, with permission.*)

and 72 percent respectively.[41] When clinical examination was combined with mammography, the accuracy increased to 92 percent. Ultrasound has been reported capable of diagnosing breast cancer in 90 percent of cases;[42] sensitivity for lesions less than 2 cm was 82 percent.

Recently, diaphanography, a refined method of transillumination, has been revived with variable results in limited clinical trials. Computerized tomography has also been employed in the study of breast diseases. Unfortunately the necessity for high contrast infusion makes the procedure a semi-invasive one.[17]

REFERENCES

1. Moskowitz M: Screening for breast cancer: How effective are our tests? A critical review. Ca-A Cancer J for Clinicians 1983; 33:26–39

2. Greenwald P, Masca PC, Lawrence CE, et al: Estimated effect of breast self-examination and routine physician examinations on breast cancer mortality. New Engl J Med 1978; 299:271–273

3. McNeil BJ, Pace PD, Gray EB, et al: Preoperative and follow-up bone scans in patients with primary carcinoma of the breast. Surg Gynecol Obstet 1978; 147:745–749

4. Front D, Schneck SO, Frankle A, et al: Bone metastases and bone pain in breast cancer. JAMA 1979; 242:1747–1748

5. Citrin DL, Hougen C, Zweibel W, et al: The use of serial bone scans in assessing response of bone metastases to systemic treatment. Cancer 1981; 47:680–685

6. Silverberg E: Cancer statistics, 1982. Ca-A Cancer J for Clinicans 1982; 32:15–42

7. Viadana E, Bross IDJ, Pickren JW: An autopsy study of some routes of dissemination of cancer of the breast. Br J Cancer 1973; 27:336–340

8. Joffe N: Metastatic involvement of the stomach secondary to breast carcinoma. Am J Roentgenol 1975; 123:512–521

9. del Regato JA: Pathways of metastatic spread of malignant tumors. Semin Oncol 1977; 4:33–38

10. Lemon HM, Wotiz HH, Parson L, et al: Reduced estriol excretion in patients with breast cancer prior to endocrine therapy. JAMA 1966; 196:1128–1136

11. Smith OW, Smith GV: Urinary estrogen profiles and etiology of breast cancer. Lancet 1970; 1:1152–1155

12. Fisher B, Ravdin RG, Ausman RK, et al: Surgical adjuvant chemotherapy in cancer of the breast: Results of a decade of cooperative investigation. Ann Surg 1968; 168:337–343

13. Whitley JE, Witcofski RL, Bolliger TT, et al: ^{99m}Tc in the visualization of neoplasms outside the brain. Am J Roentgenol 1966; 96:706–711

14. Cancroft ET, Goldsmith SJ: ^{99m}Tc pertechnetate scintigraphy as an aid in the diagnosis of breast masses. Radiol 1973; 106:441–443

15. Villarreal RL, Parkey RW, Bonte FJ: Experimental pertechnetate mammography. Radiol 1974; 111:657–662

16. Richman SD: Breast scintigraphy, in Johnston SD, Jones AE (eds): Breast Cancer Diagnosis. New York, Plenum, 1976

17. Richman SD, Brodey PA, Frankel RS, et al: Breast scintigraphy with Tc-99m pertechnetate and Ga-67 citrate. J Nucl Med 1975; 16:293–299

18. Higasi T, Nakayama Y, Murata A, et al: Clinical evaluation of Ga-67 citrate scanning. J Nucl Med 1972; 13:196–201

19. Tonami N, Matsuda H, Koizumi K, et al: Study of cases with ^{67}Ga-uptake in the breast. Radioisotopes 1980; 29:552–555

20. Larson SM, Milder MS, Johnston GS: Interpretation of the Ga-67 photoscan. J Nucl Med 1973; 14:208–213

21. Serafini AM, Raskin MM, Zand LC, et al: Radionuclide breast scanning in carcinoma of the breast. J Nucl Med 1974; 15:1149–1153

22. Telfer N, Lee YN, Merill Q, et al: Rubidium-86 uptake by red blood cells of breast cancer patients. J Surg Oncol 1981; 16:159–165

23. Soimakallio S, Kiuru A: In-11 bleomycin imaging of breast tumors. Eur J Nucl Med 1980; 5:369–371

24. Warczyglowa D, Szostak S, Marzecki Z, et al: Bleomycin labeled with ^{99m}Tc for differentiation for breast tumors. Eur J Nucl Med 1981; 6:59–61

25. Katzenellenbogen JA, Heiman DF, Goswami R, et al: Halogenated estrogen analogs as potential breast tumor imaging agents. Cancer Treat Rep 1979; 63:1214–1215

26. Ross RJ, Thompson JS, Kim K, et al: Nuclear magnetic resonance imaging and

evaluation of human breast tissue: Preliminary clinical trials. Radiol 1982; 143:195–205

27. Clark DG, Painter RW, Sziklas JJ: Indications for bone scans in preoperative evaluation of breast cancer. Am J Surg 1978; 135:667–672

28. Citrin DL: The role of the bone scan in the investigation and treatment of breast cancer. CRC Crit Rev in Diag Imaging 1980; 12:39–55

29. Hellman RS, Wilson MA: Discordance of sclerosing skeletal secondaries between sequential scintigraphy and radiographs. Clin Nucl Med 1982; 7:97–99

30. Fritz SL, Preston DF, Gallagher JH: ROC analysis of diagnostic performance in liver scintigraphy. J Nucl Med 1981; 22:121–128

31. Brennan MJ: Breast cancer, in Holland JF, Frei (eds): Cancer Medicines. Philadelphia, Lea and Febiger, 1973, p 1785

32. Lunia S, Parthasavathy KL, Bakishi S, et al: An evaluation of ^{99m}Tc colloid liver scintiscans and their usefulness in metastatic workup: A review of 1,424 studies. J Nucl Med 1975; 16:62–67

33. Davies RJ, Vernon M, Crost DN: Liver snaps and the detection of clinically unsuspected liver metastases. Lancet 1974; 1:279–281

34. Dendy PP, Keyes WI, Reid, A, et al: A clinical trial of the value of a tomographic section view to identify liver abnormalities by radionuclide imaging, with special reference to metastatic disease. Eur J Nucl Med 1981; 6:51–55

35. DeRivas L, Coombes RC, McCready VR, et al: Tests for liver metastases in breast cancer: Evaluation of liver scan and liver ultrasound. Clin Oncol 1980; 6:225–230

36. Roberts JG, Chare MJ, Leach KG, et al: Spleen size in women with breast cancer. Clin Oncol 1979; 5:317–323

37. Ege GN: Internal mammary lymphoscintigraphy in breast carcinoma: A study of 1072 patients. In J Rad Oncol Bio Phys 1977; 2:755–761

38. Deland FH, Kim EE, Corgan RL, et al: Axillary lymphoscintigraphy by radioimmunodetection of carcinoembryonic antigen in breast cancer. J Nucl Med 1979; 20:1243–1250

39. Leger JL, Naimark AP, Beique RA, et al: Report of the "Ad Hoc" Committee on Mammography. J Canada Ass Radiol 1974; 25:3–9

40. Wolfe JN: Xeroradiography of the Breast. Springfield, Illinois, Charles C. Thomas, 1972

41. Isard H, Becker W, Shilo R, et al: Breast thermography after four years and 10,000 studies. Am J Roentgenol 1972; 115:811–816

42. Kobayashi T, Takatari O, Hattori N, et al: The sensitivity graded method of ultrasonotomography and clinical evaluation of its diagnostic accuracy. Cancer 1974; 33:940–946

CHAPTER 6

Tumors of the Digestive Tract

INTRODUCTION

Cancers of the alimentary tract are often difficult to diagnose in their early stage. The main drawback to successful treatment is that most lesions do not produce symptoms until late in their course, and hence are usually advanced on initial presentation. For example, the onset of adenocarcinoma in the pancreas is insidious and initial symptoms often masquerade as vague nonspecific gastrointestinal complaints or constitutional signs. Until better diagnostic tests are available for screening and detecting small localized lesions, improvement in the rather grim survival figures of cancer involving this organ will remain unchanged.

The power of nuclear medicine to reveal disease is accomplished through the high degree of contrast with minute amounts of radiopharmaceuticals. This high contrast is the result of the differential accumulation of administered radionuclide in diseased versus nondiseased tissues. Although the availability of CT and ultrasonography (US) has reduced emphasis on some aspects of radionuclide imaging, particularly pancreas imaging, other areas exist where radionuclide imaging remains active, such as in the liver. Radionuclide liver-spleen imaging can be performed on virtually any patient without preparation, is technically easy to reproduce, and gives structural and functional information about the entire organ. At present, scintigraphy remains a useful screening procedure for metastatic malignancy, since it is very sensitive, least expensive, and is often used to guide liver biopsy and to follow the course of disease. Ultrasonography is also used in screening as it is comparable to nuclear medicine procedures in price, but has some difficulty with obese patients and bowel gas. Computed tomography is also an important procedure in searching for metastatic disease in other abdominal organs and lymph nodes with the problem of occasional contrast dye allergy.

Because of the propensity for carcinoma of the colon and rectum to metastasize to the liver, and the fact that the patient with liver metastases is frequently asymptomatic with or without abnormal liver function tests, many authorities rely on liver-spleen imaging before the treatment, and at 3 to 6 month intervals to guide therapy.

Radionuclide bone scanning may also be indicated, since as many as 17 percent of patients with colon and rectal carcinomas have been reported to have bony metastases in the absence of metastatic liver disease.[1] Those patients with suspected biliary or pancreatic tumor and a bilirubin level of below 5 mg/dl, are initially evaluated with radionuclide hepatobiliary scanning. The selenomethionine pancreas scan has, to a large extent, been replaced by CT and ultrasonography. Tumors of the gastrointestinal tract, except for primary liver cancers, are detected with low sensitivity by Ga-67 imaging. Venous thrombosis occurs commonly in cancer patients, particularly with pancreatic primary, and radionuclide venography visualizes flow patterns without pain and the risk of sensitivity reactions to contrast media. Gastrointestinal bleeding from neoplastic disease commonly arises distal to the ligament of Treitz, and positive radionuclide scans help the angiography locate the bleeding point. Brain scanning with radionuclides, though largely replaced by CT scans, can still be of use in evaluations of metastatic disease where the circumstances warrant.

CHARACTERISTICS OF DIGESTIVE TUMORS

Epidemiology

The incidence of esophageal carcinoma exhibits wide variations from one country to another. In the third national cancer survey the adjusted rates were 4.7 and 1.3 for white males and females, and 16.7 and 3.7 for black males and females, respectively.[1] In Sweden and Finland 40 to 50 percent of the patients are women. The incidence rate is highest in Japan, with a rate of 46.3 per 100,000 men over 35. This disease is common in elderly males over age 60. It may appear at a younger age in association with Barrett's esophagus with achalasia or hiatus hernia.[2] Those with Plummer-Vinson syndrome have a 91 times greater susceptibility than the general population, and lye stricture is considered important etiologically in Finland.[3] Smoking and alcohol are also important predisposing agents. There are also differences in the distribution of carcinomas possibly related to their etiology; a higher proportion of carcinomas of the upper third is reported from South Africa, whereas the mid-esophagus is the more common site in China.[4]

In the third national cancer survey, the sex-age adjusted incidence rates for cancer of the stomach were 15.1 and 7.1 for males and females. This is a disease of middle and later life with peak incidence in the 50 to 59 year age group. This disease accounts for about 10 percent of cancer deaths in the United States, but both incidence and mortality have steadily decreased in the United States in the past 40 years. The estimate of new stomach cancer cases in 1983 was 24,500.[5] The incidence of stomach cancer is highest in

Japan, and is higher in Japanese in the western sector of the United States than in other ethnic groups there. It seems to be more common in those with a high starch and low fresh fruit and vegetable intake, and is also seen with increased frequency following Billroth II gastrectomy, in some cases many years postoperative.

The small bowel is one of the largest epithelial organs in the body, but carcinomas or any other tumors of the small intestine are quite rare. The third national cancer survey revealed an age-adjusted incidence of 1.2 and 0.8 for white males and females, respectively, per 100,000 population. Crohn's disease has been considered as a precursor of cancer of the small intestine,[6] as has asbestosis.

The estimated cancer incidence in the large bowel (colon plus rectum) in the United States in 1983 is 14 and 15 percent of all cancers for males and females, and the estimated cancer deaths of the large bowel in the United States in 1983 are 12 and 15 percent of all cancers for males and females, respectively.[5] The incidence of cancer of the large bowel is lower in most countries in the world than in the United States. Two out of three patients are over 50 years of age. With familial polyposis or chronic ulcerative colitis, cancer appears at an earlier age. There is no evidence that diverticula are important in development of cancer, but high-fat diets are implicated in an increased incidence of large bowel cancer.

The estimated incidence and deaths from pancreatic cancer in the United States in 1983 is 3 and 5 percent respectively.[5] Males are affected more often than females, and the tumors occur mainly between the ages of 35 and 70 years. There is no evidence that chronic pancreatitis or alcoholism has any etiologic significance. Smoking, however, is implicated.

Carcinoma of the liver, the most frequent cancer in the world related to hepatitis B infection, is relatively rare in the United States. There is a very high incidence among Orientals and among the Bantus of South Africa. It occurs much more often in males than in females. The average age at onset is between 60 and 70 years. Cirrhosis of the liver is present in about 70 percent of patients in the United States and Europe and the relationship with hepatocellular carcinoma is with postnecrotic and posthepatic cirrhosis.[7] Cirrhosis is a much less common association in third world countries. Malignant vascular tumors, as well as hepatomas, have been observed after a long lapse in patients in whom Thorotrast has been used for diagnostic purposes.[8] Increased incidence is also associated with vinyl chloride, arsenic, and oral contraceptives.

Pathology and Metastasis

The most common appearance of the esophageal carcinoma is a fungating mass protruding into the lumen, or ulcer. Twenty percent of cases are in the upper third of the esophagus; 37 percent in the middle and 43 percent in the lower third. No less than 95 percent of esophageal tumors are squamous cell carcinomas, and adenocarcinomas seldom arise in the upper third of the esophagus. The majority of carcinomas of the gastroesophageal junction are adenocarcinomas of gastric origin.

Because the esophagus lacks a serosal covering, tumors may spread outside of it and directly invade various important structures within the chest.

In the upper third of the esophagus, dissemination through the lymphatics may lead to lymph nodes of the anterior jugular chain or of the supraclavicular region. Tumors of the middle third may metastasize to the mediastinum and also the subdiaphragmatic lymph nodes. Tumors of the lower third metastasize predominantly to abdominal lymph nodes. If the tumor is less than 5 cm, 90 percent have metastatic disease. Metastasis through the blood vessels may occur. Spread is to liver in 32 percent and to lungs or pleura in 21 percent of patients.

Practically all carcinomas of the stomach arise from mucus-secreting cells. Adenocarcinoma is the most common malignant tumor, comprising 97 percent or more of the cancers. Ulcerative carcinoma is the most common variety, and linitis plastica is an uncommon type. Among the sarcomas, 60 percent are lymphomas, and leiomyosarcomas are next in frequency. Leiomyomas are the most common benign tumors of the stomach, and they are unusual as a cause for clinical symptoms. Carcinomas of the stomach usually arise in the prepyloric and antral regions. Squamous carcinoma arising in the esophagus can invade the stomach.

Adenocarcinomas of the stomach metastasize very frequently to regional lymph nodes and the liver. Lymphangitic metastases of the lung also may be observed, and the proportion of ovarian and bone metastasis is small. Polypoid carcinomas metastasize less frequently than sessile carcinomas. Malignant lymphoma commonly involves the perigastric and adjacent retroperitoneal lymph nodes; metastases to spleen, pancreas, and liver are common. Leiomyosarcomas metastasize only rarely to regional lymph nodes, but frequently to lungs, liver, and other distant organs.

In a total of 199 tumors of the small intestine, 77 were benign and 122 malignant.[9] Adenomas, lipomas, and leiomyomas are the most common benign tumors, and they are most frequently encountered in the duodenum. Adenocarcinomas, carcinoids, and malignant lymphomas are the most common malignant tumors, and adenocarcinomas are more frequently found in the duodenum and jejunum.

Adenocarcinomas metastasize first to regional lymph nodes and sometimes to the liver, lungs, and bone. Carcinoids vary in their ability to metastasize: those arising in the ileum do so more frequently than those in the appendix. Malignant lymphomas frequently extend to the regional lymph nodes and later to distant areas. Leiomyosarcomas metastasize to the liver and lungs and infrequently to the lymph nodes.

Any lesion that causes an intraluminal protrusion in the large bowel is usually called a polyp. The most common epithelial polyps are the hyperplastic polyps, usually less than 5 mm in diameter and without a stalk (sessile). Because of the occurrence of cancer in familial polyposis, and because adenomatous polyps of the large bowel often are found together with cancer (the incidence is correlated with size), the polyps have been thought to be precancerous. The villas adenomas found predominantly in the rectosigmoid are usually single. Carcinomas usually spring from the normal mucosal of the colon, and 16 percent occur in cecum and ascending colon, 8 percent in transverse colon and splenic flexure, 6 percent in descending colon, 20 percent in sigmoid, and 50 percent in rectum. However, these statistics are changing in recent experience with higher percentages in ascending and transverse colon. With few exceptions, carcinomas of the large bowel are adenocarcinoma, with the majority of moderate or well differentiation. Carcinomas may extend through the wall of the serosal surface, where adherence to neighboring organs may occur. Duke's classification depends upon depth of anatomic spread and presence or absence of nodal metastases: (1) invasion into submucosa and muscle, (2) invasion into serosa, (3) invasion through serosa and involvement of regional nodes, and (4) distant metastases. Cecal carcinomas may directly invade the lateral abdominal gutter, and at times, the abdominal wall. Involvement of the pancreas, gallbladder, liver, spleen, and stomach may also occur.

The lymphatic spread of carcinomas of the large bowel proceeds from lymph node, progressing along the aorta as far as the mesenteric and pancreatic lymph node areas. Once the thoracic duct is reached by the tumor, metastases may appear in the supraclavicular nodes. Liver metastases are present in approximately 20 percent of patients with colon cancer at the time of diagnosis, and in 10 percent of those with rectal carcinomas.[10] At autopsy, metastases to the liver, lungs, kidneys, adrenal glands, and brain are often found; bone metastases from colon and rectal carcinoma were 11.7 and 19.4 percent, respectively.[11] There seems to be no relationship between the size of the primary lesion and the presence or absence of metastases, but undifferentiated tumors do metastasize more frequently. The higher the lesion is located and the deeper the infiltration of the wall, the greater the relative chance of visceral metastases.[12]

Most of the pancreatic carcinomas arise from ductal epithelium, and adenocarcinomas comprise the majority of all malignant tumors of the pancreas. Overall, carcinoma of the pancreatic head makes up about two-thirds, and carcinoma of the tail about one-fourth of all cases.

Carcinomas of the pancreatic head have a tendency to remain relatively localized, whereas those of the body and tail metastasize widely. Tumors in the pancreatic head may invade as well as compress the common bile duct and portal veins. They may directly infiltrate the musculature of the duodenum, stomach, and transverse colon. Tumors of the pancreatic body and tail may directly invade the spleen, left kidney, adrenal gland, and diaphragm. Single or multiple venous thromboses frequently are associated with these tumors. Islet cell tumors are uncommon, and most are functioning tumors. Ten percent of the functioning tumors are multiple, and about 10 percent are malignant. The tumors are distributed evenly through the head, body, and tail of the pancreas. The presence of metastases is an indication of malignancy.

Cystadenomas and cystadenocarcinomas are exceedingly rare, slowly growing tumors located usually in the body and tail of the pancreas. These tumors are multiloculated and usually benign. Metastases of pancreatic carcinomas frequently involve the regional lymph nodes, and blood-borne liver metastases are usually multiple. Metastases to the lung and bones also are commonly observed. Islet cell carcinomas often metastasize to the liver, and less frequently to the lymph nodes.

Most malignant tumors of the liver are adenocarcinomas. There are two major cell types: (1) those arising from the liver cell (hepatoma), and (2) those arising from the bile duct cell (cholangiocarcinomas). The tumor can present as a large primary lesion. The right lobe is more frequently involved than the left. There is usually nodularity of the noninvolved liver parenchyma in associated cirrhosis. Venous collateral circulation, ascites, esophageal varices, and splenomegaly may be present.

Because of the tendency of primary hepatic carcinoma to invade the hepatic and portal veins, the tumor may migrate to the heart and lungs. Other sites of metastases are the brain, kidney, and regional lymph nodes. Cholangiocellular carcinomas may metastasize to bones.[13]

Clinical Evolution

The early symptoms of esophageal carcinoma are dysphagia, a frequent symptom that is eventually present in almost all patients, and weight loss, eventually present in about 75 percent of all patients. Other signs and symptoms include malaise, anorexia, hematemesis, and cervical adenopathy. Ominous findings include left vocal cord paralysis from recurrent nerve palsy, and persistent cough from fistulae into the trachea. Exsanguination may occur due to communication with a large vessel such as the aorta.

Vague epigastric discomfort not responding to routine medical management with or without anorexia and weight loss, must be viewed as possible carcinoma of the stomach. The ulcer type of pain occurs much less frequently, but again when ulcers occur carcinoma of the stomach must be considered and endoscopy with biopsies performed. Iron deficiency anemia without obvious cause suggests the possibility of a lesion in the gastrointestinal tract, and gastric cancer is one of these. Unfortunately, many gastric cancers produce no symptoms at all until after metastases occur. Ascites may occur, and the tumor may metastasize to lymph nodes around the bile ducts and cause jaundice.

Benign tumor of the small intestine is the most frequent cause for intussusception in adults. Carcinomas obstruct by napkin-ring constriction. Lymphomas frequently obstruct by disturbing bowel motility through invasion and nerve destruction rather than by producing actual stenosis. Perforation is an uncommon presenting finding usually indicative of lymphoma or sarcoma. Bleeding occurs in over half of the symptomatic patients, and abdominal mass may be present with or without other symptoms. Malabsorption syndrome with sprue-like picture occurs with lymphomas, but is rare. There is an increasing number of cancer syndromes associated with small intestinal tumors. A carcinoid syndrome occurs in about 20 percent of all patients with carcinoid tumors metastatic to the liver. It is characterized by cutaneous flushes and cyanosis, chronic diarrhea and peristaltic rushes, respiratory distress, and vulvular disease of the right heart. Malignant lymphomas involving the duodenum or upper portion of the jejunum are associated with diarrhea and steatorrhea.

Patients with carcinoma of the large bowel often disregard the early symptoms. Symptoms depend upon the location of tumor in colon; right colonic lesions are typically present with unexplained anemia, and tumors of the sigmoid usually present with obstruction from napkin-ring growth. Villous adenoma of the rectum is a rare cause of profound electrolyte depletion from major loss of mucus into the bowel lumen from tumor surface.

Most patients with pancreatic cancer present with insidious onset of depression, asthenia, anorexia, weight loss, gaseousness, and nausea. Loss of weight is the most common symptom and typically is rapid and marked. Pain occurs in 70 to 80 percent of patients, and is usually dull or boring, and confined to the epigastrium or back. With carcinoma in the pancreatic head, the above symptoms usually precede the slow and progressive development of jaundice, which is usually persistent and accompanied by pruritis. The gallbladder may dilate and become palpable. With carcinoma in the pancreatic body, there is infringement upon the abundant nerve plexus around the celiac ganglia, which results in back pain that is made worse by eating and by lying down. With carcinoma in the pancreatic tail, the initial symptoms are frequently caused by metastases to other organs in association with severe weight loss and weakness. Symptoms of functioning islets cell tumors are due to the over-production of insulin and other hormones. With slight hypoglycemia, there is fatigue, lassitude, restlessness, and malaise.

The initial symptoms of hepatic cancer are nonspecific complaints of weakness, anorexia, abdominal fullness or bloating, and dull aching upper abdominal pain. With progressive growth, the pain becomes more severe and radiates into the back. The liver becomes palpably enlarged and tender, and often one can palpate tumor nodules on its surface. Ascites, which often obscures the weight loss, is almost inevitable. Clinical signs and symptoms of portal hypertension can be present. Jaundice may range from mild to severe.

NUCLEAR IMAGING

Radionuclide Esophagus and Stomach Dynamic Swallow Imaging

In general, esophageal and stomach imaging following swallowing of a bolus of radionuclide labeled material is of little value as a screening method in tumor lesions, but may be of help in determining the extent of a lesion. With extrinsic obstruction, an accumulation of radionuclide occurs at the level of the obstruction and the transit time is prolonged in proportion to the narrowing. Radionuclide tests may reveal esophageal or stomach fistulas before they are radiographically apparent. The test may be repeated as frequently as needed, and is therefore useful as a means to follow therapy.

Tumor-associated gastroparesis is more common than appreciated and it can be seen with pancreatic cancer without causing obstruction or anatomical invasion of the stomach. Gastric emptying study with Tc-99m DTPA and metoclopramide is helpful in the palliative care of gastrointestinal tumor patients.[14]

Radionuclide Pancreatic Imaging

Despite 16 years of use, the pancreas scan with selenomethionine is still not generally accepted as an efficacious procedure because of poor diagnostic image quality due to factors such as concentration of radionuclide in the overlying liver and surrounding viscera as well as varying degrees of Se-75 concentration in the pancreas. When used, it is usually reserved for patients in whom no abnormality has been seen on ultrasound, computerized tomography, or endoscopic retrograde cholangiopancreatography (ERCP), and the suspicion of pancreatic disease remains high. A normal pancreas scan rules out pancreatic disease with 90 to 95 percent certainty, but an abnormal study

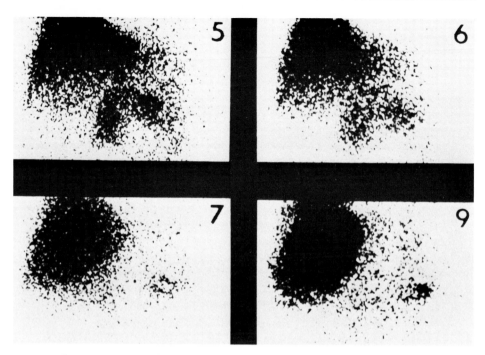

Figure 1. Longitudinal tomographic images of upper abdomen with Se-75 selenomethionine shows good visualization of pancreatic head, body, and tail.

has less certain significance.[15] The most common scan abnormality in pancreatic carcinoma is decreased uptake throughout the pancreas. Therefore, it is impossible to differentiate between pancreatitis and carcinoma of the pancreas by scan alone. The appearance of solitary or multiple filling defects within the pancreas is more common in carcinomas than in pancreatitis. Small tumors less than 2 cm in diameter cannot be detected, but patients referred for pancreas scanning usually have symptoms that occur only at a late stage in the course of pancreas carcinoma. Functioning islet-cell tumors associated with Zollinger-Ellison syndrome and their metastases have been found to show increased Se-75 uptake.[16] Some evidence suggests that iodinated diphenylhydantoin may act as a specific imaging agent for insulinoma.[17]

Multiplane tomographic scanning (Fig. 1) offers improved spatial resolution in the pancreas at different depths, and tomographic scans have been judged superior to regular camera images in ten of 20 cases and inferior in three.[18]

Scintigraphy with selenomethionine after glucose load also enhanced the uptake of the radioactivity in the pancreas and improved the differential diagnosis between pancreatitis and pancreatic tumors.[19] The substitution of carbon-11 for carbon-12 in amino acids inspires hope for improved pancreatic concentration of C-11 valine, C-11 tryptophan, and C-11 aminocyclobutane carboxylic acid. Of 14 abnormal positron emission computerized tomographic scans, eight had the pancreatic carcinoma, three abdominal lymphoma, and three pancreatitis.[20] Resolution and contrast of emission computerized tomography were superior to conventional scans, but inferior to ultrasound and CT.

Radionuclide Liver Imaging

Although continuing technological changes make consensus difficult to establish on the choice of an imaging procedure for the liver, the strengths of radionuclide liver imaging are that it is simple to perform, all patients are examinable, the procedure is reasonably sensitive, the waiting time for results is minimal, it can provide information about reticuloendothelial function, and the cost is moderate. The weaknesses are that spatial resolution is only adequate for deep lesions, there is poor delineation of lesions adjacent to the liver, and it is usually ineffective for characterizing lesions as to their etiology.[21] At present, scintigraphy remains a useful diagnostic procedure for suspected biliary tract and liver disease, and computerized tomography is especially valuable in distinguishing extrinsic from intrinsic hepatic lesions when ultrasound results are equivocal. The sensitivities for the metastatic disease to the liver were CT 96 percent, scintigraphy 94 percent, and ultrasound 77 percent. Specificities were 86 percent, scintigraphy 67 percent, and ultrasound 50 percent.[22]

In recent years, there have been a number of approaches to improve the results from radionuclide imaging of the liver: minimizing motion artifacts using the upright position, breath holding, and respiratory gating; using emission tomography; and combining results of liver scanning with liver function tests or CEA.[23] Scintiangiography is often added to the standard static liver imaging, but its usefulness in differentiating benign from malignant disease is limited.[24] However, computerized flow scintigrams providing the arterialization index raised the diagnostic sensitivity for metastatic liver disease from 85 to 100 percent in 228 patients.[25] Different pathologic patterns (encapsulated tumors or diffuse parenchymal involvement) of hepatoma leads to variable patterns on liver scanning with sulfur colloid, and the combination of defects on colloid imaging, with increased gallium uptake, suggests hepatoma, abscess, or—rarely—metastatic disease. Benign tumors of the liver (Fig. 2) come to the attention of nuclear medicine far less often. Liver cell adenoma and focal nodular hyperplasia are pathologically distinct: one of the differences is the presence of bile ducts and Kupffer cells in focal nodular hyperplasia, whereas both are absent in liver cell adenoma. The radiocolloid scan usually detects these lesions when they are clinically significant in size. A mass showing uptake of sulfur colloid is more likely to be focal nodular hyperplasia.[26] Depending on their size and histologic nature, hemangiomas involving the liver may be manifest as focal defects. Recently, delayed imaging with a blood pool agent (Tc-99m RBC) has been recommended as an immediate sequel to radiocolloid imaging to display the delayed filling so characteristic of hemangiomas (Fig. 3) in the liver.[27]

Some primary tumors, such as colorectal carcinoma and renal cell carcinoma, appear to generate metastases that rapidly lead to focal defects (Figs. 4 and 5). Tc-99m sulfur colloid hepatic scintigraphy showed 96 percent sensitivity and 92 percent specificity for the metastatic lesion (Figs. 6 and 7) in 92 patients with colon carcinoma.[28] Sixty-three percent of patients with abdominal lymph node involvement had a lesion on liver scan, but CEA levels were normal in 58 percent of patients. Preoperative liver scan identified metastatic lesions (Figs. 8 and 9) in six out of seven patients with suspected gastric cancer.[28] Other tumors such as cancer of the lung or breast

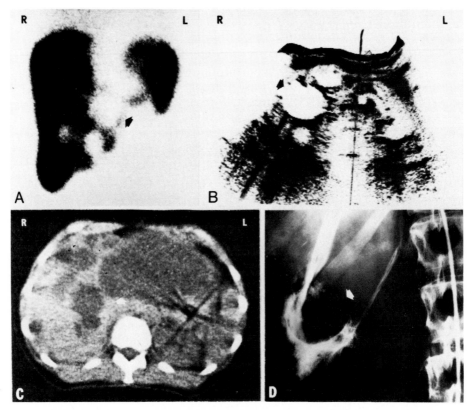

Figure 2. Polycysts in the liver demonstrated by Tc-99m sulfur colloid imaging (**A**), ultrasonography (**B**), CT scan (**C**), and contrast angiography (**D**).

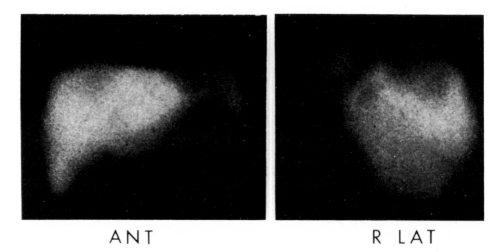

Figure 3. Anterior and right lateral views of the liver with Tc-99m sulfur colloid show defects in the upper portion of right lobe. At operation a hemangioma was found.

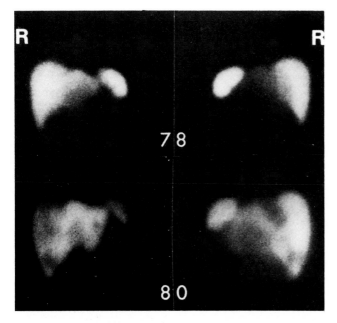

Figure 4. Anterior and posterior views of liver-spleen with Tc-99m sulfur colloid show multiple metastatic lesions of colon cancer in both lobes of the liver developed in 2 years.

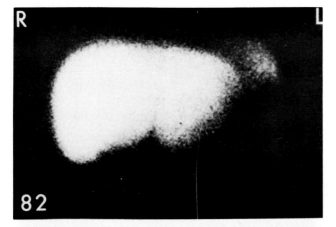

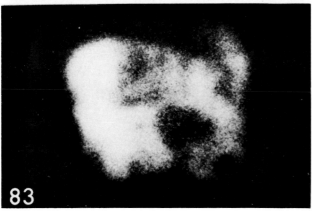

Figure 5. Anterior views of liver-spleen colloid scan show multiple metastatic lesions of esophageal cancer developed in one year.

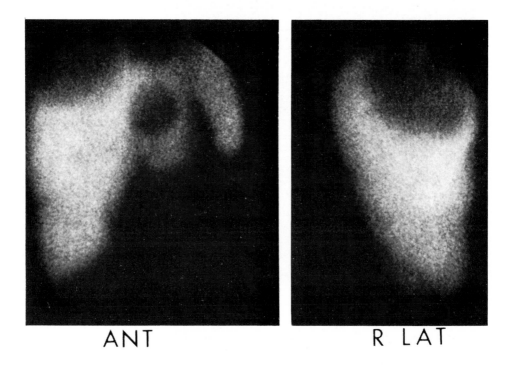

ANT   R LAT

Figure 6. Anterior and right lateral views of the liver with Tc-99m sulfur colloid shows a large metastatic lesion of colon cancer in the upper portion of right lobe. Also noted is a cyst in the left lobe which is well defined on the CT.

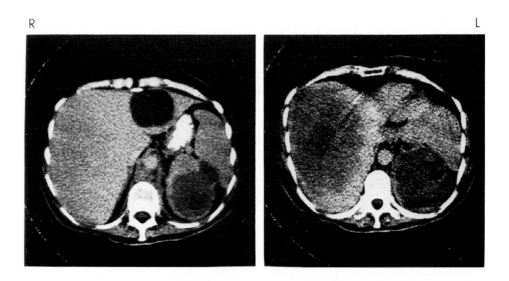

Figure 7. Right lateral view of same liver as in Fig. 6.

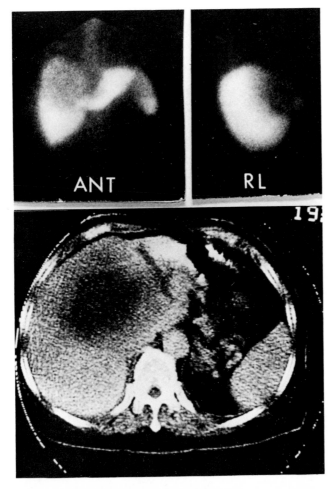

Figure 8. Anterior and right lateral views of the liver with Tc-99m sulfur colloid (upper row) show a large metastatic lesion of gastric leiomyosarcoma. CT also reveals a metastasis.

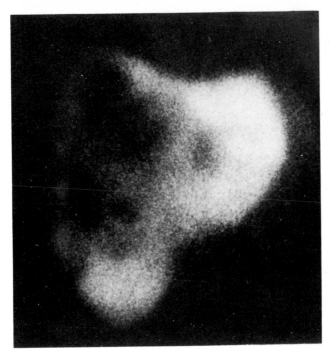

Figure 9. Anterior hepatic image of colloid liver scan shows multiple metastatic lesions of gastric leiomyosarcoma in the right lobe.

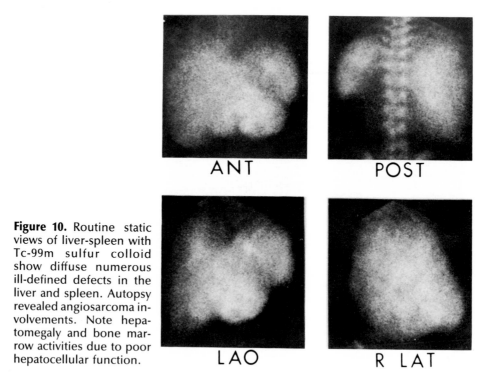

ANT POST

LAO R LAT

Figure 10. Routine static views of liver-spleen with Tc-99m sulfur colloid show diffuse numerous ill-defined defects in the liver and spleen. Autopsy revealed angiosarcoma involvements. Note hepatomegaly and bone marrow activities due to poor hepatocellular function.

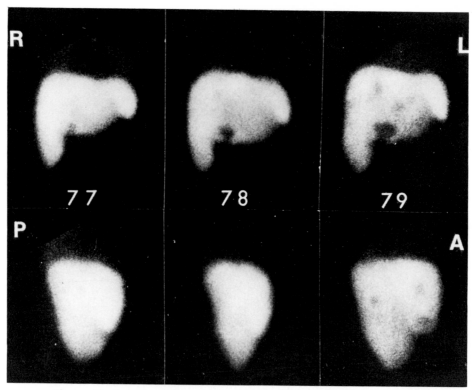

Figure 11. Anterior (upper row) and right lateral (lower row) views of the liver with Tc-99m sulfur colloid show progressive development of carcinoid metastases in both lobes.

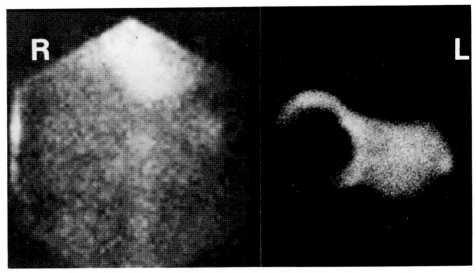

Figure 12. Anterior blood-pool (left) and static (right) images of liver-spleen using Tc-99m sulfur colloid show a large hepatoma in the right hepatic lobe in a nonalcoholic patient.

are often associated with hepatic metastases that are widespread and are manifest as diffuse heterogeneity of radiocolloid distribution (Figs. 10 and 11).

A hepatoma (Figs. 12–14) presented as a focal defect on sulfur colloid scan and was well visualized with hepatobiliary imaging. Biliary agents may help to differentiate hepatomas from hemangiomas or gallium avid metastatic tumors.[28] Cancerous and noncancerous causes of biliary tract obstruction produced significantly different findings in hepatobiliary imaging of 11

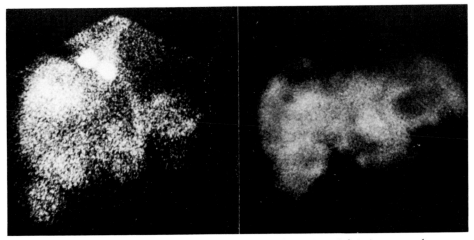

Figure 13. Anterior blood-pool (left) and static (right) images of colloid liver-spleen show multicentric hepatomas in an alcoholic patient. Note some vascularity of lesions.

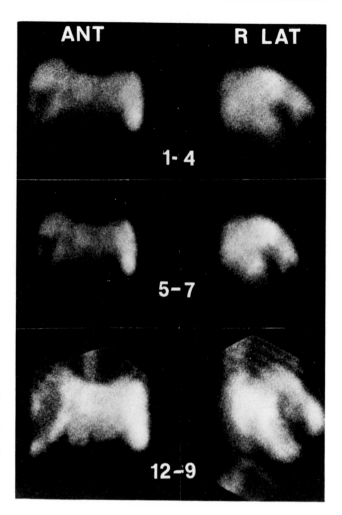

Figure 14. Anterior and right lateral views of follow-up colloid liver-spleen scans in an alcoholic patient show multicentric hepatomas and progressive deterioration of hepatic function.

patients with extrahepatic biliary obstruction secondary to cancer; 7 out of 9 (78 percent) had complete obstruction and 9 of 11 (82 percent) had a moderate to severe decrease in hepatocyte clearance.[29] In suspected small biliary duct obstruction, the best strategy is radionuclide scan first, followed by ultrasonography; and in suspected large duct obstruction, ultrasonography first, followed by radionuclide scan.[30] Using Tc-99m labeled hepatobiliary agents, the need for laparoscopic biopsy was obviated in a majority of patients by identifying the gallbladder fossa shown as occupying the discrete defect on sulfur colloid scans.[31]

Liver subtraction scan using Se-75 selenomethionine (Fig. 15) in 16 hepatoma patients gave a true positive rate of 89 percent. Of the 40 patients who did not have hepatomas, 32 scans showed no evidence of selective concentration of selenomethionine, giving a true negative rate of 80 percent.[32]

There have been reports of abnormal accumulation of Tc-99m phosphate compounds (Figs. 16 and 17) or F-18 bone scanning agent in hepatic metastases from mucinous colonic adenocarcinoma, well differentiated adenocarcinoma of the lung, and stomach adenocarcinoma.[33,34]

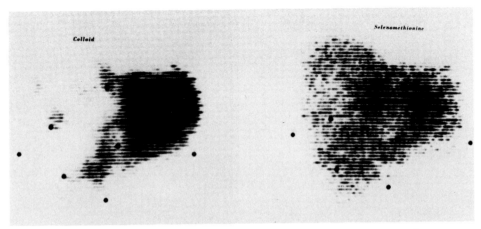

Figure 15. Tc-99m sulfur colloid scan shows multiple defects in the right hepatic lobe. Selenomethionine scan shows uptake of radioactivity in the lesions. Histology revealed a well differentiated hepatoma. *(From Br J Radiol 1980; 53:539, with permission.)*

The use of Tc-99m macroaggregated albumin (MAA) infusion studies in intrahepatic arterial chemotherapy offers an excellent evaluation of catheter placement and tumor perfusion, in addition to helping avoid gastrointestinal complications.[35] Three different patterns of tumor perfusion (Fig. 18) were observed in 184 MAA studies: increased central radioactivity (33 percent); decreased central radioactivity (33 percent); and mixed and/or diffuse radioactivity (33 percent). Tumor arteriovenous shunting was demonstrated in 38 percent—showing localization of radiotracer activity in the lungs—and decreased as tumors decreased in size.

Radionuclide Spleen Imaging

Tc-99m sulfur colloid scan is an effective method to evaluate spleen size, position, and focal or diffuse alterations. Sonography can be useful as the next step to determine if focal defects are solid or cystic, and to show the relationship of the spleen to adjacent organs. Of 900 patients undergoing radionuclide liver-spleen imaging, the spleen was abnormal in 65 patients, 41 with splenomegaly and 24 with focal defects. Splenomegaly was secondary to liver disease in 39 patients and to leukemia in 2 patients. Focal defects (Fig. 19) were caused by lymphoma, metastatic carcinoma, lymphangiomatosis, multiple myeloma, hematoma, and abscess.[36]

Disorders that may cause a reversal of the liver-to-spleen uptake ratio on sulfur colloid scans include hepatic parenchymal disease, congestive heart failure, and chemotherapy. Forty-seven of 147 melanoma patients were reported to have increased splenic uptake.[37] Relatively decreased splenic uptake of sulfur colloid was noted in patients with pancreatic cancer,[38] possibly related to splenic vascular alteration, but the exact pathophysiology is not known.

Accessory spleens may appear as left upper quadrant masses on sulfur colloid scans and may be difficult to differentiate from liver tissue. Heat-damaged Tc-99m RBC may be used to visualize the spleen without interference from the liver.

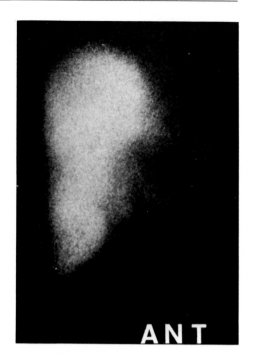

Figure 16. Anterior colloid hepatic image shows almost complete replacement of left hepatic lobe by metastatic colon cancer.

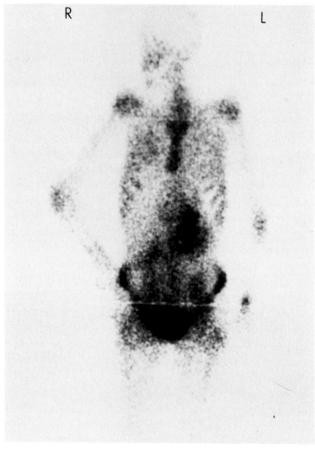

Figure 17. Anterior whole-body bone scan shows an uptake of Tc-99m MDP in the metastatic colon cancer. See legend for Fig. 16.

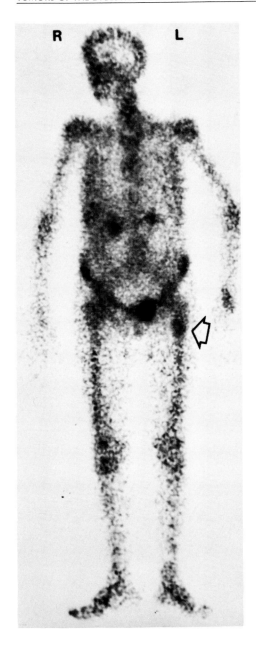

Figure 21. Anterior whole-body image of MDP bone scan shows a metastatic lesion (arrow) of esophageal cancer in left proximal femur.

useful for planning radiotherapy in patients with esophageal cancer because missing extraesophageal extension or gross metastases from radiation fields were detected.[41] Of 38 clinical stage II patients, 15 (39 percent) could be stage III by the results of gallium scan.

Gallium scan showed an intense uptake in both flanks extending to the subdiaphragmatic region for diffuse infiltrating peritoneal mesothelioma in a patient with ascites demonstrated by CT and ultrasonography.[42]

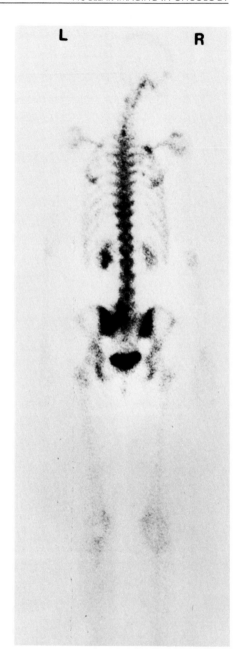

Figure 22. Posterior image of whole-body bone scan with Tc-99m MDP shows metastatic lesions of colon cancer in left side of L_5 and S_1 bodies.

Radionuclide Lymphangiography

Tc-99m antimony sulfide colloid is rapidly removed by lymphatic flow after interstitial injection, and appears useful in patients with obstruction to lymphatic drainage from metastatic involvement. It is possible that the imaging of abdominal lymph nodes may be improved with tomographic devices.

Lymphoscintigraphy was performed prior to surgery by injection of Tc-99m rhenium sulfur colloid into the submucosal layer above and below esophageal carcinomas. Metastases were found in 34.6 percent of visualized

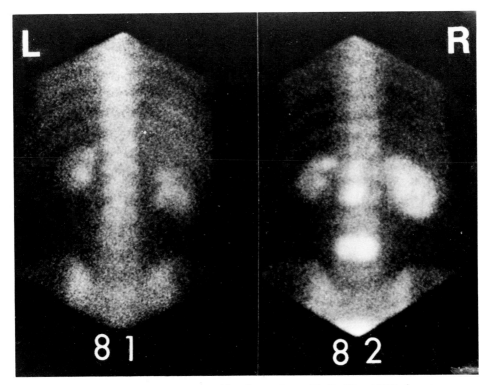

Figure 23. Posterior images of lumbar spine with Tc-99m MDP show developed metastatic lesions of colon cancer in L_2, L_3, L_4 bodies in 1 year.

and 3.8 percent of nonvisualized nodes.[43] Therefore, lymphoscintigraphy improves the preoperative evaluation of mediastinal node involvement, with hot nodes indicating an increased probability of metastases.

Radionuclide Study for Gastrointestinal Bleeding and Venous Thrombosis

Gastrointestinal bleeding from neoplastic disease most commonly arises distal to the ligament of Treitz, and bleeding from the upper tract in patients with solid tumors is commonly caused by hemorrhagic gastritis, peptic ulcer, esophagitis, or varices. In the diagnosis of lower tract bleeding, both colonoscopy and barium enema are compromised in the presence of active bleeding. Angiography for the diagnosis and treatment of lower tract bleeding can be more precisely directed if preceded by a radionuclide study (Fig. 27) using Tc-99m sulfur colloid or RBC.[44]

Venous thrombosis occurs commonly in cancer patients especially with pancreas primaries, and tumor recurrences in the pelvis can often be confused with lower extremity venous thrombi. Radionuclide venography visualizes flow patterns as with contrast venography, but without pain and the risk of sensitivity reactions to contrast media (Fig. 28). It has been useful in evaluating the need for prophylactic therapy in patients who have experienced or are at risk of developing deep vein thrombosis, and in monitoring therapeutic regimens.[45]

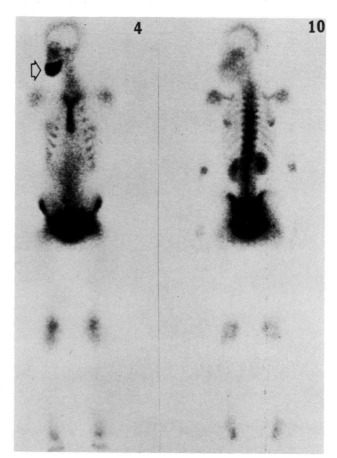

Figure 24. Longitudinal sectional images of whole-body with Tc-99m EDP show a metastatic lesion of rectal carcinoma in the mandible (arrow).

Other Nuclear Studies

Yttrium-90 Microsphere Therapy. Twenty-five patients with metastases to the liver from colorectal cancer received Yttrium-90 microspheres and 5-fluorouracil on a continuing basis through the intra-arterial catheter, and they survived an average of 26 months.[46] The dose of 100 mCi of Yt-90 was well tolerated by the liver.

Radioimmunodetection. Using a high-titer, purified goat anti-CEA IgG labeled with I-131, the sensitivity of the radioimmunodetection of colorectal cancers was found to be 90 percent in 27 patients.[47] Very high circulating CEA titers did not appear to hinder successful tumor radiolocalization. Neither the one case of a villous adenoma receiving the anti-CEA IgG nor the two cases of colonic cancer receiving normal goat IgG showed tumor localization. This method may be of value in preoperatively determining the location and extent of disease, in assessing possible recurrence or spread postoperatively, and in localizing the source of CEA production in patients with rising

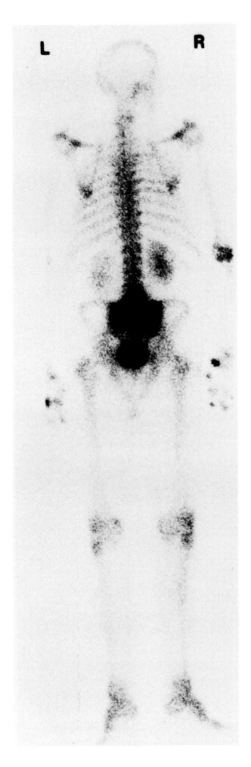

Figure 25. Posterior whole-body scan image with Tc-99m EDP shows a metastatic lesion of rectal carcinoma in the sacrum.

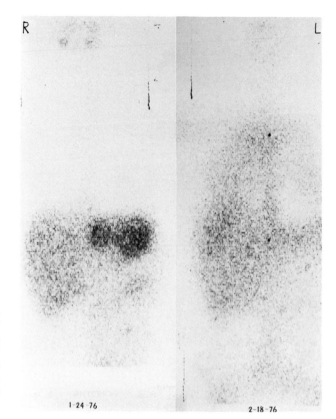

Figure 26. Anterior whole-body images using Ga-67 citrate show Burkitt's lymphoma involvement in left hepatic lobe and spleen (left), and complete resolution by chemotherapy in 25 days.

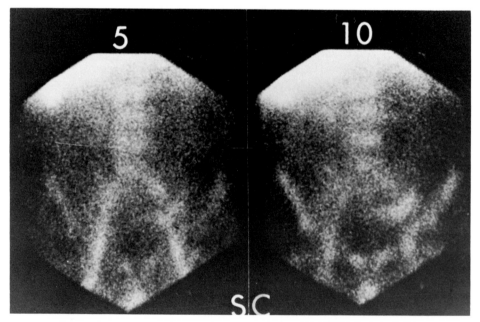

Figure 27. Anterior abdominal images at 5 and 10 minutes following the injection of Tc-99m sulfur colloid show an extravasated activity (active bleeding) in the sigmoid in a patient with colon cancer.

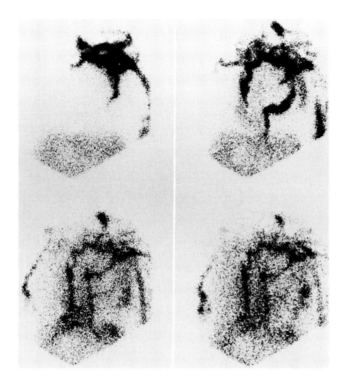

Figure 28. Anterior serial images of the chest following the injection of Tc-99m sulfur colloid into left antecubital vein show an obstruction of left subclavian vein with multiple collateral circulations in a patient with carcinoid metastases in the left upper lung.

or elevated CEA titers. An ancillary benefit could be a more tumor-specific detection test for confirming the findings of more conventional diagnostic measures.

Positron Emission Tomography Imaging. Positron emission tomography is a new imaging technique with the potential for noninvasive measurement of tissue metabolism in living patients.

The experience to date indicates that C-11 tryptophan images are technically superior to Se-75 selenomethionine images,[48] and they can define both morphological and functional alterations in the pancreas. All hepatic metastases demonstrated by Tc-99m sulfur colloid scan in 10 patients were easily identified and generally were seen more clearly by PET imaging using Ga-68 iron hydroxide colloid.[49] In one patient, additional lesions that were not identified on the initial sulfur colloid scan were detected.

Nuclear Magnetic Resonance Imaging. The clinical applications of NMR imaging are rapidly gaining in scope and significance. Studies dealing with the anatomy and pathology in various organs show great promise for this modality. Focal disease within the liver was demonstrated by NMR and CT imagings, although an area of focal atrophy and another hepatic infarction were only recognized with NMR.[50] NMR easily differentiates malignant tumors from benign cystic lesions and provides useful information in patients with cirrhosis and metastatic deposits. NMR was also accurate in the demonstration of pancreatic pathology in 12 patients.[51]

OTHER RADIOLOGIC PROCEDURES

Computed Tomography

A major advantage of CT is its ability, with higher spatial resolution, to display in cross-section the contour and the substance of the organ. In the diagnosis of space-occupying lesions of the liver, CT may be used either as the initial method of detection or as a complementary method to confirm the presence or clarify the nature of a lesion that has been detected by another imaging modality.

Most lesions detected by CT, regardless of their nature, are of a diminished radiographic density. Because of greater resolution and because images are composed of relatively thin sections of tissue, CT scans show smaller lesions than Tc-99m sulfur colloid scans.

The diagnosis of pancreatic carcinoma by CT is based upon gross focal alterations in size and shape of the gland. The tissue density of the pancreatic carcinoma is similar to that of normal pancreatic parenchyma. Peripheral calcification has been described in islet cell tumor and cystadenoma. Tumor mass originating from the posterior surface of the left lobe of the liver or other peripancreatic area can be confused with primary pancreatic carcinoma. Besides detecting abnormalities of the liver, spleen, pancreas, or bone, CT can demonstrate lymphadenopathy. The accuracy of CT was virtually identical to lymphangiogram for the detection of para-aortic lymph node involvement. CT is also an excellent means for guiding biopsy procedures for hepatic, pancreatic, or retroperitoneal lesions, precluding the necessity for laparotomy to provide histologic diagnosis of disease.

Ultrasonography

Advantages of US over CT are shorter scan time, ability to scan in longitudinal and oblique planes, lower cost, and no ionizing radiation. Solid masses are seen as localized or diffuse disruptions of the normal acoustic pattern of the liver. Ultrasound images are degraded to a much greater degree in large patients than are CT scans.

Primary tumors of the pancreas—most commonly adenocarcinoma—appear as irregular and often lobular structures that, on US, have fewer internal echoes than normal pancreatic parenchyma. Neither US nor CT is reliable in the differentiation of benign or malignant solid masses without the aid of ancillary signs such as liver or lymph node metastases. Whereas US has the theoretical advantage in thin patients in differentiating pancreatic carcinoma from adenopathy in the parapancreatic region because the latter exhibits a more homogenous echo pattern than that of the pancreas, US has more unsuccessful examinations due to overlying bowel gas which precludes delineation of the pancreas.

Angiography

Increasing availability and technical advances of abdominal CT and US have caused a decline in the use of diagnostic angiography and a shift of emphasis toward interventional angiography. Nonetheless, angiography has retained its position as the principle diagnostic examination of conditions caused by vascular disease, or in which vascular conditions may dictate management. While CT has become the principle technique for staging tumors and preoperative assessment of extension through fascial planes or into adjacent organs, super-selective arteriography is often called upon to meet the need of radical surgery for detailed preoperative mapping of the vascular supply. Angiographically, hepatomas, and hepatocellular carcinomas are characterized by neovascularity and a tumor stain persisting throughout the early venous phase. Carcinoma of the pancreas can be suggested on the basis of arterial encasement and characteristic venous occlusions.

Selective and superselective catheterization of organ vessels has been exploited for infusion of pharmacologic agents, chemotherapeutic perfusion, perfusion with chilled solutions for the purpose of creating a hypothermic milieu, and for transcatheter embolization with inert embolic material or radioactive infarct particles.

Contrast Radiographic Studies

Fluoroscopy is paramount for the patient with suspected carcinoma of the esophagus. The typical cone-shaped dilatation surmounting an area of stenosis, irregularity, and displacement, is usually that of an advanced lesion. Cineradiography may detect fleeting abnormalities that might be undetected in routine fluoroscopy.

Fluoroscopic studies of mucosal patterns, and of the motility and pliability of stomach walls, are very important for the evaluation of stomach cancer. The early radiographic signs may be limited to changes in the mucosal pattern. The leather-bottle shape of the stomach in linitis plastica is typical. The radiographic diagnosis of small bowel tumors is also made on the basis of alterations in the mucosal pattern and the presence of obstruction or extraluminal filling defects. Carcinomas of the large bowel are usually visualized as an intraluminal filling defect breaking the smooth continuity of the bowel wall. Sessile polyps showing a dimpling or indentation of the colonic wall often prove to be malignant. The most frequent changes of the pancreas cancer observed in radiographic examination are deformity, narrowing, ulceration, and filling defects of the duodenum. A pancreatic tumor may become sufficiently larger to cause obvious widening of the duodenal loop.

Endoscopic Studies

Endoscopy has established itself as an important and exciting modality in the diagnosis of cancers of the alimentary tract. By incorporation of the principles of fiberoptics, endoscopes feature excellent illumination, flexibility, and superb optical imaging. They also offer the opportunity to obtain biopsies.

REFERENCES

1. Cutler SJ, Young JL, Jr: Third national cancer survey: Incidence data. Natl Cancer Inst Monogr 1975; 41:10–27, 100–135, 388–427
2. Wychulis AR, Woolam GL, Anderson HA, et al: Achalasia and carcinoma of the esophagus. JAMA 1971; 215:1638–1641
3. Gerami S, Booth A, Pate JW: Carcinoma of the esophagus engrafted on lye stricture. Chest 1971; 59:226–227
4. Ming SC: Tumors of the esophagus and stomach, in Atlas of Tumor Pathology, second series, Fasc. 7, AFIP, Washington D.C., 1973

5. Silverberg E: Cancer statistics, 1983. Ca-A Cancer J Clinicans 1983; 33:9–25
6. Saed W, Kim S, Burch BH: Development of carcinoma in regional enteritis. Arch Surg 1974; 108:376–379
7. Mori W, Shah M: A comparative geopathological study of liver cirrhosis and primary hepatic cancer between Cambridge and Tokyo. Gann 1972; 63:765–771
8. Kuisk H, Sanchez JS, Mizuno NS: Colloidal thorium dioxide in radiology with emphasis on hepatic cancerogenesis. Am J Roent 1967; 99:463–475
9. Ostermiller W, Joergenson EJ, Weibel L: A clinical review of tumors of the small bowel. Am J Surg 1966; 111:403–410
10. Nielsen J, Balsleo I, Fenger HJ, et al: Carcinoma of the rectum with liver metastases. Acta Chir Scand 1973; 139:479–481
11. Berge TH, Ekelund G, Mellner C, et al: Carcinoma of the colon and rectum in a defined population. Acta Chir Scand 1973; 438:1–86
12. Dionne L: The pattern of blood-borne metastasis from carcinoma of rectum. Cancer 1965; 18:775–781
13. Al-Sarraf M, Kithier K, Vaitkevicius VK: Primary liver cancer. Cancer 1974; 33:574–582
14. Shivshanker K, Bennetts RW, Haynie TP: Tumor related gastroparesis: Correlation with metoclopramide. Gastroenterol 1981; 80:1283–1286
15. Partain CL, Staab EV, McCartney WH: Multiple imaging modalities for the study of pancreatic disease. Semin Nucl Med 1979; 9:36–42
16. Staab EV, Babb OA, Klatte EC, et al: Pancreatic radionuclide imaging using electronic subtraction technique. Radiol 1971; 99:633–635
17. Balachandran S, Beierwaltes WH, Ice RD, et al: Tissue distribution of ^{14}C-, ^{125}I- and ^{131}I-diphemylhydantioin in the toadfish, rat and human with insulinomas. J Nucl Med 1975; 16:775–777
18. Gaston EL, Gaspard CL, Brooks AC, et al: Multiplane tomographic imaging of the pancreas. J Nucl Med Technol 1980; 8:28–29
19. Zubovskii GA, Vasil'chenko SA: Pancreatoscintography after glucose load. Med Radiol 1982; 27:39–42
20. Buonocore E, Hubner KF: Positron emission computed tomography of the pancreas: A preliminary study. Radiol 1979; 133:195–201
21. Grossman ZD, Thomas FD: Position of nuclear imaging in the age of transmission computed tomography and ultrasound, in Freeman LM, Weissman HS (eds): Nuclear Medicine Annual 1980. New York, Raven Press, 1980, pp 367–391
22. Snow JH, Jr, Goldstein HM, Wallace S: Comparison of scintigraphy, sonography, and computed tomography in the evaluation of hepatic neoplasms. Am J Roent 1979; 132:915–918
23. Freitas JE, Dworkin HJ: Optimizing the detection of hepatic metastases. N Nucl Med 1979; 20:264–265
24. Echevarria RA, Bonnano C: Value of routine abdominal nuclide angiography as part of liver scan. Clin Nucl Med 1979; 4:66–78
25. Sarper R, Fajman WA, Tarcan YA, et al: Enhanced detection of metastatic liver disease by computerized flow scintigrams. J Nucl Med 1981; 22:318–321
26. Sandler MA, Petrocelli RD, Marks D, et al: Ultrasonic features and radionuclide correlation in liver cell adenoma and focal nodular hyperplasia. Radiol 1980; 135:393–397
27. Front D, Royal HD, Israel O, et al: Scintigraphy of hepatic hemangiomas: The value of Tc-99m labeled RBC. J Nucl Med 1981; 22:684–687
28. Utz JA, Lull RJ, Anderson JH, et al: Hepatoma visualization with Tc-99m pyriodoxylidene glutamate. J Nucl Med 1980; 21:747–749
29. Klingensmith WC, Kuni CC, Fritzberg AR: Cholescintigraphy in extra-hepatic biliary obstruction. Am J Roent 1982; 139:65–70

30. Klingensmith WC, Johnson ML, Kuni CC, et al: Complementary role of Tc-99m diethyl-IDA and ultrasound in large and small duct biliary tract obstruction. Radiol 1981; 138:177–184

31. Rao BK, Pastakia B, Lieberman LM: Evaluation of focal defects on Tc-99m sulfur colloid scans with new hepatobiliary agents. Radiol 1980; 136:497–499

32. Coakley AJ, Wraight EP; Selenomethionine liver scanning in the diagnosis of hepatoma. Br J Radiol 1980; 53:538–543

33. Kim EE, Domstad PA, Choy YC, et al: Accumulation of Tc-99m phosphate complexes in metastatic lesions from colon and lung carcinomas. Eur J Nucl Med 1980; 5:299–301

34. Tonami N, Maeda T, Aburano T, et al: Marginal accumulation of ^{99m}Tc-methylene diphosphonate in liver metastasis, from stomach carcinoma. Eur J Nucl Med 1981; 6:43–45

35. Bledin AG, Kantarjian HM, Kim EE, et al: ^{99m}Tc-labeled macroaggregated albumin in intrahepatic arterial chemotherapy. Am J Roentgenol 1982; 139:711–715

36. Shirkhoda A, McCartney WH, Staab EV, et al: Imaging of the spleen: A proposed algorithm. Am J Roentgenol 1980; 135:195–198

37. Sober A, Jr, Mintzis MM, Lew RA, et al: The significance of augmented radiocolloid uptake by the spleen in patients with malignant melanoma. J Nucl Med 1979; 20:1232–1236

38. Tatum JL, Burke TS, Fratkin MJ, et al: Relative decreased splenic uptake of Tc-99m sulfur colloid in patients with pancreatic cancer. Clin Nucl Med 1982; 7:1–4

39. Hoffer P: Status of gallium-67 in tumor detection. J Nucl Med 1980; 21:394–398

40. Waxman AD, Richmond R, Siemsen JK, et al: Correlation of contrast angiography and histologic pattern with gallium uptake in primary liver cell carcinoma—Noncorrelation with alpha-feto protein. J Nucl Med 1980; 21:324–327

41. Kondo M, Ando M, Kosuda S, et al: Ga-67 scan in patients with intrathoracic esophageal cancer planned for surgery. Cancer 1982; 49:1031–1034

42. Dach J, Patel N, Patel S, et al: Peritoneal mesothelioma: CT, sonography and gallium-67 scan. Am J Roentgenol 1980; 135:614–616

43. Terui S, Kato H, Hirashima T, et al: An evaluation of the mediastinal lymphoscintigram for carcinoma of the esophagus studies with ^{99m}Tc rhenium sulfur colloid. Eur J Nucl Med 1982; 7:99–101

44. Smith RK, Arterburn G: Detection and localization of gastrointestinal bleeding using Tc-99m pyrophosphate in vivo labeled red blood cells. Clin Nucl Med 1980; 5:55–60

45. DeNardo SJ: Role of nuclear medicine in the detection of venous thrombosis, in Freeman LM, Weissman HS (eds): Nuclear Medicine Annual 1980. New York, Raven Press, 1980, pp 341–365

46. Ariel IM, Padula G: Treatment of asymptomatic metastatic cancer to the liver from primary colon and rectal cancer by the intra-arterial administration of chemotherapy and radioactive isotopes. J Surg Oncol 1982; 20:151–156

47. Kim EE, Deland FH, Casper S, et al: Radioimmunodetection of colorectal cancer. Cancer 1980; 45:1243–1247

48. Kirchner PT, Ryan J, Zalutsky M, et al: Positron emission tomography for the evaluation of pancreatic disease. Semin Nucl Med 1980; 10:374–391

49. Kumar B, Miller TR, Siegel BA, et al: Positron tomographic imaging of the liver: Ga-68 iron hydroxide colloid. Am J Roentgenol 1981; 136:685–690

50. Doyle FH, Pennock JM, Banks LM, et al: Nuclear magnetic resonance imaging of the liver: Initial experience. Am J Roentgenol 1982; 138:193–200

51. Smith FW, Reid A, Hutchison JM, et al: Nuclear magnetic resonance imaging of the pancreas. Radiol 1982; 142:677–680

CHAPTER 7

Genitourinary Cancer

INTRODUCTION

Since no tumor-specific radiopharmaceutical has been found, the evaluation of patients with genitourinary cancer with radionuclides uses indirect methods. An assessment of the vascular characteristics of the suspected lesion, and correlation of the result with the other available clinical and radiological data, allows for some differentiation of renal lesion.

Renal imaging with Tc-99m DTPA or glucoheptonate (GH) has been used to identify and differentiate congenital or acquired renal pseudotumors from abnormal renal masses, since renal pseudotumors are composed of functioning nephrons.

Hypernephroma (Figs. 1 and 2) is often demonstrated as a hypervascular blush on dynamic flow study, and corresponds in location to a filling defect on the static image using Tc-99m DTPA, while pseudotumor reveals normal uptake of radionuclide. Thus, the renal scan (Fig. 3) can be of value in helping to diagnose anatomical variants, the importance of which lies in the avoidance of unnecessary invasive studies, surgical exploration, or nephrectomies.

The role of radionuclide renogram in cancer patients appears mainly to assess the functional status of each kidney and to demonstrate urinary obstruction.

For any patient with malignant disease to receive the most appropriate treatment, it is necessary to demonstrate metastatic lesions as accurately as possible, and this is particularly true of cancers of the kidney, bladder, prostate, and testis. Radionuclide bone and liver scans have been used for initial staging, and for subsequent follow-up of therapeutic response. Ga-67 citrate imaging has been useful in evaluating patients with testicular seminoma.

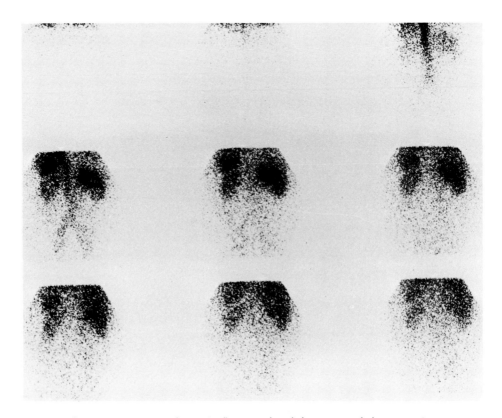

Figure 1. Posterior dynamic flow study of the upper abdomen using Tc-99m DTPA shows a focal area of increased vascularity in the upper medial aspect of the right kidney.

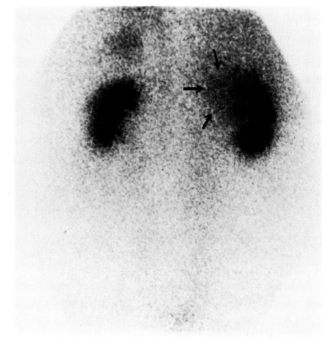

Figure 2. Static image reveals decreased radioactivity (arrows) in the same lesion as Fig. 1. Surgery confirmed a hypernephroma.

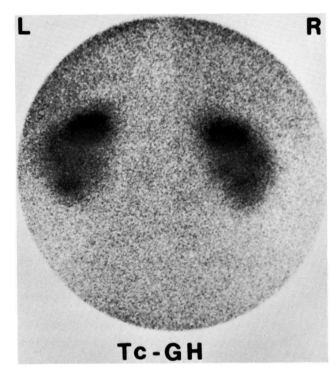

Figure 3. Posterior static image of upper abdomen with Tc-99m glucoheptonate shows relatively greater activity in upper poles of both kidneys, suggesting hypertrophied parenchymas. IVP showed questionable mass lesions in the same areas.

CHARACTERISTICS OF GENITOURINARY CANCER

Epidemiology

Renal tumors constitute slightly less than 2 percent of all malignant tumors in the United States, but make up more than 10 percent of those found in children under 5 years of age.[1] Adenocarcinomas make up about 80 percent of all malignant renal tumors and occur more frequently in men than in women, mostly over 60 years of age. Transitional cell carcinomas of the renal pelvis constitute 6 to 12 percent of all malignant renal tumors, and are found mostly in elderly patients.[2] Wilm's tumors are found equally in girls and boys, mostly under 4 years of age.[3]

Cancer of the urinary bladder constitutes 6.4 percent of all cancer in males and 2.3 percent of cancer in females in the United States.[1] The age-specific incidence was 25.2 per 100,000 for individuals 55 to 59 years old, and rose to 61 per 100,000 for individuals aged 65 to 69, and 103.2 per 100,000 for those aged 75 to 79.

Carcinoma of the prostate makes up 16.3 percent of all cancer occurring in North American males, second in incidence only to cancer of the lung.[1] Of all male cancer in the United States, it is estimated that incidence and deaths in 1982 were 18 and 10 percent respectively.[4] This form of cancer is extremely rare in young adults.

Malignant tumors of the testes constitute 1 percent of all cancer in North American males.[1] Testicular tumors are rare in North American black men. The occurrence of testicular tumors in patients with cryptorchidism is notable; 37 cryptorchid-associated tumors in a collected series of 529 patients with testicular tumors.[5]

Pathology and Metastasis

Benign renal tumors are fairly common, but because they are usually asymptomatic most are commonly observed incidentally at operation or necropsy. Such benign renal tumors include cortical adenomas, hamartomas of the renal medulla, and subcapsular leiomyomas and lipomas; and are important as they may be difficult to differentiate from malignant tumors.

Renal adenocarcinomas vary in size from 2 cm to over 10 cm, and they rarely may be bilateral. Frequently, however, they metastasize to the lymph nodes of the renal pedicle and immediately adjacent para-aortic lymph nodes. Renal veins and vena cava involvement results in lung metastases, and paravertebral vein plexus involvement results in bony metastases. Metastasis, apparently blood-borne, have been observed on the skin, heart, breasts, larynx, and eyes.[6]

Transitional cell carcinoma of the kidney have the notorious character of being accompanied by satellite lesions of the ureter and of the opposite renal pelvis and ureter. They most often metastasize to regional lymph nodes, liver, lungs, brain, and bones.[7]

Squamous cell carcinomas arising from the renal pelvis are uncommon and are frequently associated with nephrolithiasis and pyelonephritis. They are often found in regional lymph nodes. Wilm's tumors probably arise from embryonal nephrogenic tissue, and are usually large and globular. Large tumors become attached to contiguous organs by inflammatory or neoplastic adhesions. They spread toward the pelvis, invading the renal veins and, rarely, the vena cava. They spread to regional lymph nodes (60 percent of the patients) and occasionally to bones of the pelvis, spine, and ribs. Pulmonary metastases are likely to appear within 15 months of the diagnosis.[3,8] Very rarely the tumors may directly invade the small and large bowel, liver, and vertebrae.

A papilloma of the bladder cannot be distinguished grossly from a papillary, grade I, transitional cell carcinoma. The higher the degree of the carcinoma, the greater the likelihood of bladder wall invasion.

Squamous cell carcinomas constitute only about 3 percent of the bladder carcinoma, and they are deeply ulcerated, involving the musculature of the bladder.[9] Sarcoma botryoides, a rare form of malignant bladder tumor, usually occurs in young children. Because carcinoma of the bladder is often fatal due to concomitant infection including pyelonephritis before the tumor has had an opportunity to metastasize, autopsy studies reveal a variable proportion of lymph node involvement. In 52 of 89 patients with direct extravesical invasion, 33 had regional lymph node involvement (most frequently at the bifurcation of the iliac artery), 26 had liver metastases, and 18 had lung metastasis.[10]

The carcinoma-bearing prostate is not always enlarged. About one-half of all patients with carcinoma of the prostate also present nodular hyperplasia.

Capsular involvement of prostatic cancer is frequent, and the outer layer of the capsule as well as the muscle prevents further spread of the cancer. The spread of tumor to the rectum is blocked by Denonvilliers' fascia. Carcinoma may infiltrate the seminal vesicles, but seldom affects the urethra; invasion of the bladder occurs late.

Pelvic lymph node involvement is usually found first in the hypogastric nodes, then in the obturator and iliac nodes. Extrapelvic lymph node metastases are found in the para-aortic and mediastinal nodes. Visceral metastases are seen most frequently in the kidneys, adrenal glands, liver, and lungs. At autopsy, the majority (about 80 percent) of cases show osseous metastasis along the distribution of the vertebral vein plexus.[11]

About 60 percent of the testicular tumors have been considered to be pure tumors (mature and immature teratomas, embryonal carcinomas, choriocarcinomas, and seminomas), and the remainder were mixed types.[12] Seminomas, the most frequent of testicular tumors, are rare before 16 years of age; their frequency rises to a maximum between 40 and 44 years of age. Embryonal carcinomas are the most common testicular tumors in children. Choriocarcinomas occur in young adults mostly in the third or fourth decade of life. The frequency of teratocarcinomas is about the same for the first six decades of life. Yolk sac tumor (orchioblastoma; endodermal sinus; adenocarcinoma; infantile embryonal carcinoma) is a distinctive testicular tumor of childhood.

Ipsilateral external iliac lymph node metastases occur frequently in testicular tumors. About 80 percent of the patients with embryonal carcinomas, 75 percent with teratocarcinomas, and more than 50 percent of those with seminomas and mixed tumors present retroperitoneal metastases.[13] Seminomas are the least aggressive, and contralateral metastases have been observed only from tumors of the right testis.[14] Spread of testicular tumors may take place through the spermatic veins either to the renal vein on the left or the vena cava on the right. All germinal tumors may metastasize to the liver and lungs, and also to the adrenal glands, kidneys, spleen, brain, bones, bowel, and pancreas.[15]

Clinical Manifestation

Renal cell carcinomas may be the unsuspected primary lesion of a variety of metastatic manifestations. Since metastasis may occur early, the symptoms and signs produced may mimic and suggest a diversity of clinical disorders. Hematuria is the most common symptom of renal tumors, and is the first manifestation in about 50 percent of the patients. Pain is the first symptom in about one-third, and a palpable mass completes the triad of the most common signs and symptoms. Carcinomas of the renal pelvis manifest themselves initially in most patients by painless but profuse bleeding. The most frequent first sign of Wilm's tumor is a palpable mass. As the tumor grows to involve the renal capsule or nerves, pain appears. Anemia is common, but hematuria is infrequent.

Hematuria is the most common symptom of bladder carcinoma and is prominent in carcinomas with a low degree of malignancy, whereas in the more malignant varieties, probably because of invasion of the bladder wall, there is a preponderance of pain. Since a high percentage of the tumors grow in close relation to the ureter, urinary infection is common and produces fever, weight loss, and costovertebral tenderness.

The initial symptoms of a large group of patients with prostatic carcinoma are frequency of urination, difficult or painful urination, and pain possibly due to perineural space involvement. Lumbar pain is frequently caused by osseous metastasis or urinary infection. The usual cause of death of patients with incurable prostatic cancer is renal insufficiency complicated by infection, but a number die of unrelated causes such as cardiovascular and cerebrovascular accidents.

Tumors of the testes often develop slowly and insidiously. Painless swelling is a common complaint. Lumbar pain may result from the development of retroperitoneal metastases, and symptoms of urinary obstruction may result from the compression of the renal pedicle. Gynecomastia has been observed in patients with choriocarcinoma, and virilization may be observed in children with interstitial cell tumors.

Testicular tumors may spread rapidly and result in death within 8 months of diagnosis, and lung metastasis may cause considerable symptomatology. Brain metastases are not uncommon.

NUCLEAR IMAGING

Imaging of Primary Tumors

Radionuclide renal imaging has been used to evaluate suspected renal masses, renal size, shape, position, and function (Fig. 4). Forty patients with excretory urographic findings indicating a possible, but not definite, mass lesion, were studied. Scintigraphy correctly identified 17 true masses and 17 normal variants. Four false positive and two false negative results were obtained.[16]

Sty et al.[17] reported that renal scintigraphy with Tc-99m glucoheptonate is more accurate than excretory urography in demonstrating minimal renal function and in detecting focal abnormalities. Multicystic kidneys can be distinguished from hamartomas and Wilm's tumors with the use of Tc-99m imaging.

In a few cases, renal tumors have been reported to exhibit positive uptake of radionuclide. Tc-99m DMSA uptake in renal cell carcinoma and angiomyolipoma was seen in the early image, and then an area of decreased activity was observed in the late image.[18] A Tc-99m GH renal scan also showed two functioning kidneys of markedly different size, but no focal defects in the mesoblastic nephroma. The kidney was infiltrated by a mass composed of fibrous and mesenchymal stroma surrounding islands of normal glomeruli and tubules on histopathology.[19]

Although most uptake of gallium-67 in the renal area is related to infection, occasionally it is due to tumors including renal carcinoma, lymphoma, leukemia, melanoma, and sarcoma.[20]

In a study of 22 patients having renal visualization on gallium-67 scans, 5 patients had tumors involving the kidney at autopsy, while others have shown radiographic or laboratory evidence of tumor in the kidney.[21] There was a report of abnormal Ga-67 uptake in renal cell carcinoma that also showed a hot spot on early renal imaging with Tc-99m DMSA.[22] However, bladder and prostate tumors are not well identified on gallium scans, which have been found to be of little value in these tumors.

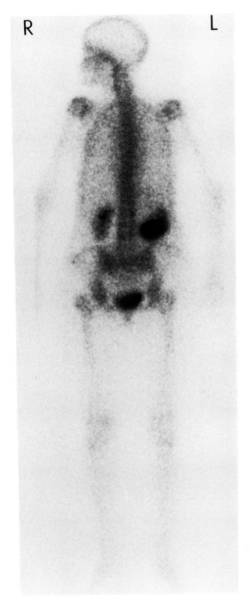

Figure 4. Longitudinal tomographic whole-body image with Tc-99m MDP shows normal bony structure, but decreased activity in the upper pole of left kidney with laterally deviated axis. Surgery found a renal cell carcinoma in the upper pole of the left kidney.

Whole-body imaging using Ga-67 (Fig. 5) has been reported as showing tumor localization in patients with seminoma of testis.[23] Scrotal imaging with Tc-99m sodium pertechnetate has been useful in differential diagnosis in patients presenting with either scrotal mass or painful scrotum.

The use of Zn-69m for prostatic cancer detection has been studied in ten men with prostatic disease, but it was useless as a diagnostic tool because of poor quality of images.[24]

A total of 15 scans using I-131 labeled antibody to alpha-feto-protein (AFP) were performed on 12 patients, ten with malignant teratoma, one with an endodermal sinus tumor, and one with a seminoma (as negative control).

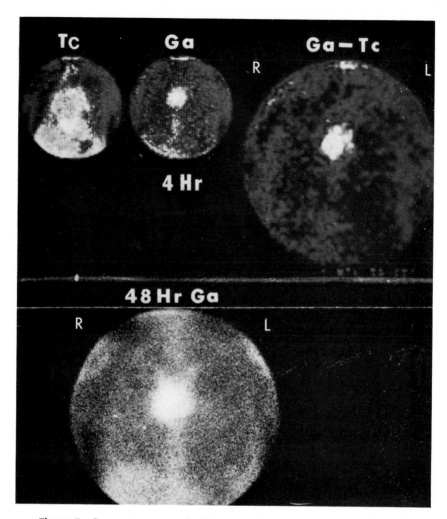

Figure 5. Computer-assisted subtraction image (upper row) of anterior chest using Ga-67 citrate and Tc-99m HSA shows a primary seminoma which is also outlined on routine 48 hour Ga image (lower row).

All nine patients with elevated serum AFP gave positive scans. Three patients gave normal scans when studied again after successful treatment and when the serum AFP had fallen to normal.[25]

Imaging of Metastatic Tumors

It is of paramount importance in the investigation of prostate cancer to detect bony metastases that usually form osteoblastic lesions due to a considerable amount of new bone formation. It has been clearly shown that the radionuclide bone scan (Figs. 6 to 9) is a highly sensitive means of detecting skeletal metastasis in patients with prostatic cancer.[26] In 28 out of 31 cases, the bone scan revealed metastases that were not visible on concomitant radiographs but that were radiographically confirmed 1 year later.[27]

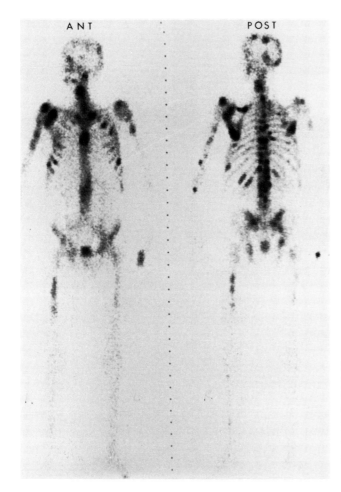

Figure 6. Anterior and posterior whole-body images with Tc-99m MDP show multiple metastatic lesions (hot spots) of prostate cancer in skull, ribs, scapulae, vertebrae, pelvis, humeri, and femora.

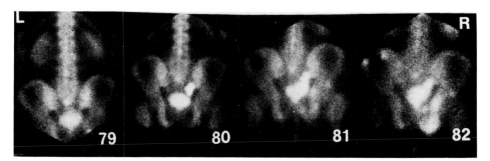

Figure 7. Posterior static images of lower spine and pelvis show progressive metastatic lesions of prostate cancer in right sacrum, left iliac, and right pubic bones on follow-up bone scans with Tc-99m MDP.

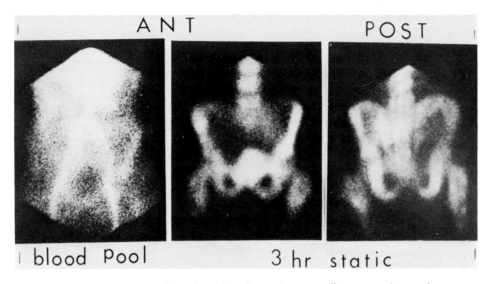

Figure 8. Anterior blood-pool and anterior as well as posterior static pelvic images with Tc-99m MDP show a large cold metastatic lesion of prostate cancer in right iliac bone in addition to diffuse metastases (hot lesions) in pelvis, lower lumbar vertebrae, and femora.

Widespread bone metastases (Fig. 10) can occasionally give rise to a uniform distribution of Tc-99m MDP resulting in a superficially normal appearance of the bone scan. The scans are recognizable by the high ratio of bone to soft tissue activity and the absence of focal lesions in the axial skeleton, and there are usually no renal images (Figs. 11 and 12). These

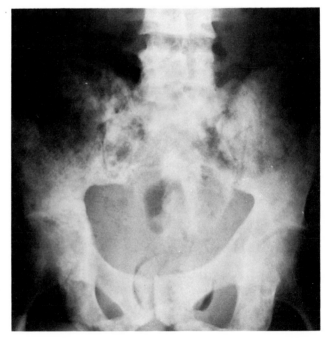

Figure 9. Pelvic radiograph shows diffuse osteoblastic lesions and possible osteolytic lesion in right iliac bone.

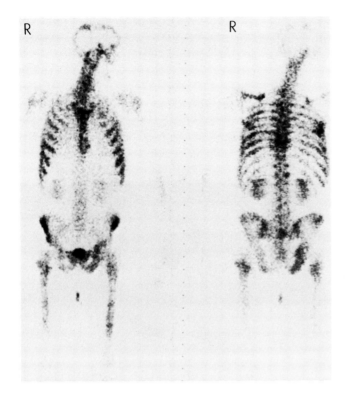

Figure 10. Anterior and posterior whole-body images using Tc-99m MDP show disseminated metastases of testicular choriocarcinoma in skull, ribs, spine, pelvis, and femora.

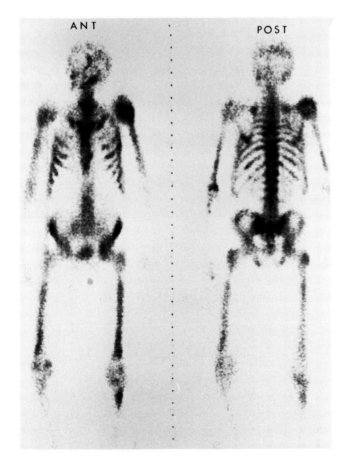

Figure 11. Anterior and posterior whole-body images with Tc-99m MDP show diffuse metastases of prostate carcinomas in all bones producing "super-scans." No significant renal activity is noted.

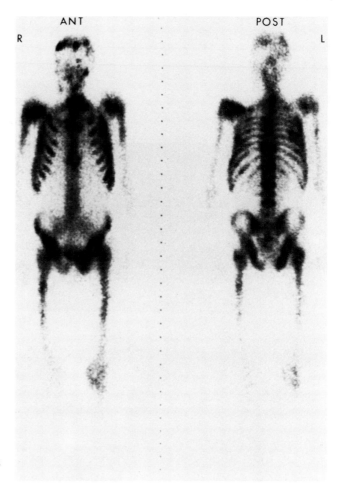

Figure 12. Similar case as Fig. 11.

"super-scans" are thought to be more frequently associated with prostatic carcinoma than with other etiologies.[28] Serial bone scans were compared with skeletal x-ray surveys and acid phosphatase in the follow-up of 60 patients with prostatic carcinoma. In this study the bone scan proved to be a more accurate monitor than the other indices.[29] In prostatic cancer patients who responded well to diethylstibesterol therapy, bone scan showed dramatic improvement that usually corresponded with clinical improvement.[30]

Bone metastasis from Wilm's tumor is a rare entity. Confirming the diagnosis was difficult in three patients because scintigraphy yielded variable results with increased (hot lesion) or decreased (cold lesion) activities.[31]

Bone scanning (Figs. 13 to 16) may be of some value in patients with renal or bladder cancer, but not to the same extent as prostatic cancer. It will be useful in the preoperative workup of the patient being considered for nephrectomy or cystectomy.

The liver scan generally can be of considerable help in metastatic disease (Figs. 17 to 20). However, it has been found that routine liver scanning before treatment is not warranted in patients with malignant genitourinary tumors, due to low predictive value and benefit:cost ratio.[32,33]

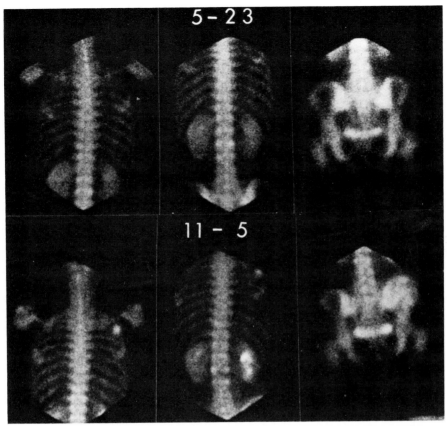

Figure 13. Posterior static images of spine and pelvis with Tc-99m MDP show developed metastases of renal cell carcinoma in right ilium and scapula (hot lesions). Also noted is a cold lesion in L$_{2-3}$ on the right in 6 months.

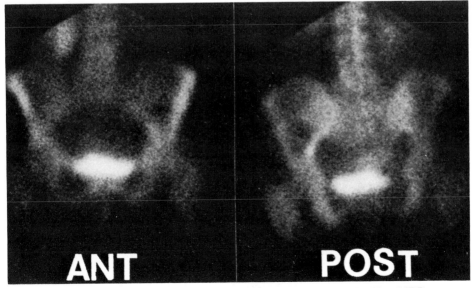

Figure 14. Anterior and posterior pelvic images using Tc-99m MDP show a cold metastatic lesion of renal cell carcinoma in left iliac bones.

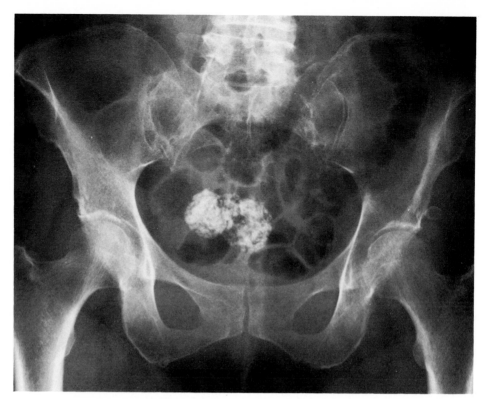

Figure 15. Same lesion as in Fig. 14. Pelvic radiograph also shows osteolytic lesion in left ilium.

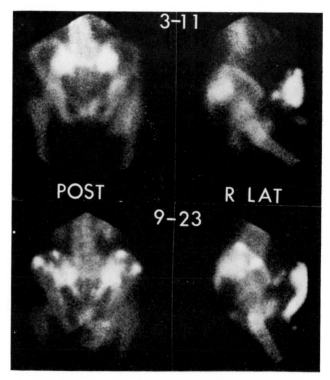

Figure 16. Posterior and right lateral views of the pelvis with Tc-99m MDP show progressive metastases of bladder carcinoma in iliac bones bilaterally.

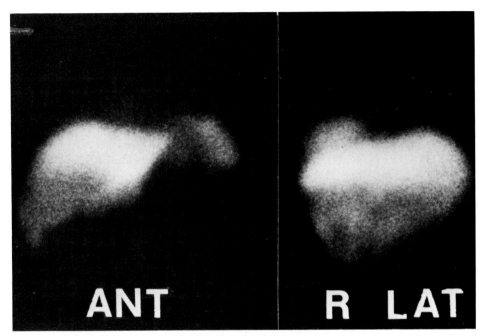

Figure 17. Anterior and right lateral view of liver-spleen with Tc-99m sulfur colloid show multiple metastatic lesions of renal carcinoma in the inferior portion of the right hepatic lobe.

One area in which gallium scan has been helpful in the past is in the staging of testicular seminoma or embryonal carcinoma (Fig. 21) and prostatic carcinoma (Fig. 22). Testicular tumors have some similarity to lymphomas in that they are both radiosensitive and radiocurable, if the entire extent of the disease is treated. The gallium scan was accurate in staging 43 of 46 patients, and more valuable than lymphangiogram in evaluating abdominal involve-

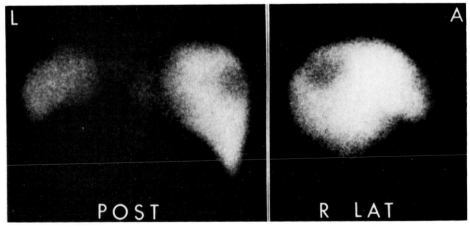

Figure 18. Posterior and right lateral views of colloid liver-spleen scan show a metastatic lesion of bladder carcinoma in the upper portion of right heptic lobe posteriorly.

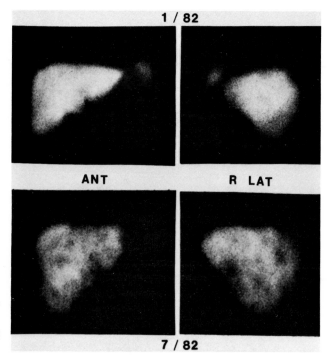

Figure 19. Anterior and right lateral colloid liver-spleen images show progressive metastases of prostate carcinoma in right and left hepatic lobes in 6 months.

ment with testicular tumor.[34] However, recently its use in this situation is limited due to the availability of other, better imaging techniques.

Silent metastases to regional lymph nodes are frequent in the early stages of prostatic carcinoma. Gardiner et al,[35] developed and improved a simple method for demonstrating prostatic lymphatic drainage by a simple median injection of Tc-99m antimony sulfide colloid into the prostatic capsule. Ege[36] also demonstrated effectively the groups of nodes constituting the internal iliac or pelvic lymphatics by bilateral perianal injection of Tc-99m antimony sulfide colloid. The simplicity of this method suggests potential for pre- and intra-operative identification of the lymph nodes (Fig. 23).

Radioimmunodetection using I-131 labeled antibodies to AFP and human chorionic gonadotropin (hCG) permitted detection and localization of AFP or hCG-producing tumors (Fig. 24).

Other Nuclear Procedures

Radionuclide angiography with Tc-99m DTPA proved to be a suitable noninvasive method for the follow-up of patients with renal cell carcinomas, undergoing palliative arterial embolization. If radionuclide angiogram revealed residual perfusion of the tumor, functional scintigraphy with I-131 Hippuran was performed to quantify the residual function.[39]

Radiation nephritis can lead to reduced function, renal hypoplasia, and hypertension. These findings can be evaluated by renal imaging and renography. Such studies can be of value in renal localization and shielding during radiotherapy to avoid renal damage.

The Ga-67 scan appears to be useful for early detection of lung toxicity

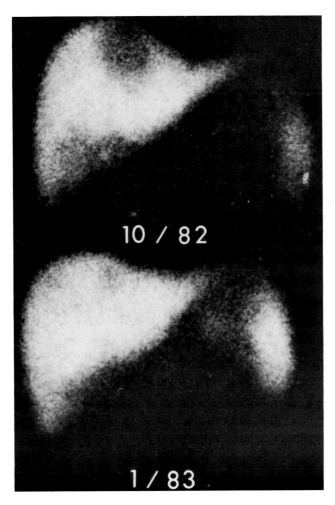

Figure 20. Anterior colloid liver-spleen images show significantly resolving metastatic lesion of testicular carcinoma 3 months following chemotherapy.

after bleomycin therapy. Widespread uniform uptake of Ga-67 over both lung fields has been observed in a patient with testicular teratoma after he received more than 600 mg of bleomycin. The patient had no respiratory symptoms and his chest x-rays were normal. Then prednisone was started and a scan repeated 11 days later showed that the lung fields had returned to normal.[40]

OTHER RADIOLOGIC IMAGING

Conventional radiography of the abdomen delineates the renal contour and nonspecific masses. It is useful in the detection of calcification and fat within the mass. Nephrotomography, when technically optimal, will define a renal cyst with high accuracy.

An intravenous pyelogram (IVP) is performed to detect a mass distorting the pelvocalyceal system or bladder, and obstructing or deviating the ureters. Retrograde pyelography may provide information not available by IVP, particularly in lesions of the pelvocalyceal system.

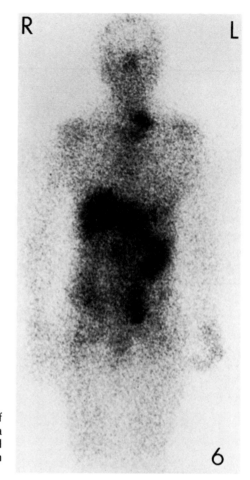

Figure 21. Longitudinal section image of whole body using Ga-67 citrate shows a metastatic lesion of testicular embryonal carcinoma in left supraclavicular lymph nodes.

Cystography is used to determine the extent of a bladder carcinoma, and cystoscopic examination of all bladder tumors is essential. When a renal mass is identified, ultrasonography is helpful to differentiate benign cysts from neoplasms. Computed tomography can often define the nature of the primary neoplasm and also demonstrate tumor invasion to adjacent viscera or retroperitoneal lymph nodes.

Angiography should be undertaken if the diagnosis is still in doubt. The vascularity of a metastases will mimic that of the primary neoplasm. Spermatic vein phlebography, which can be done at the time of orchiectomy, may be useful in demonstrating metastatic lymph nodes.

Contrast lymphangiography has been useful in patients with prostate or testicular tumors, providing valuable information as to the abdominal and pelvic lymph node involvement.

A chest radiograph is always helpful for detecting pulmonary metastasis and pleural effusion. Limited skeletal x-rays for the questionable lesions on bone scans are essential to enhance diagnostic specificity.

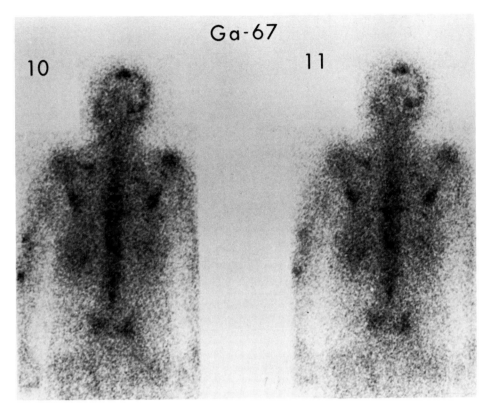

Figure 22. Longitudinal tomographic body images with Ga-67 citrate show multiple metastatic lesions of prostatic cancer in skull, ribs, scapulae, and vertebrae.

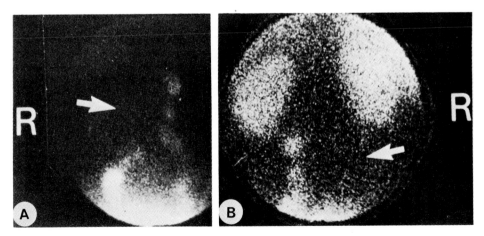

Figure 23. Anterior (**a**) and posterior (**b**) views 3 hours following Tc-99m antimony colloid injection show complete absence of radiocolloid uptake in right para-aortic nodes (arrows). Microscopic examination revealed testicular cancer metastasis. *(From Radiol 1983; 147:231, with permission.)*

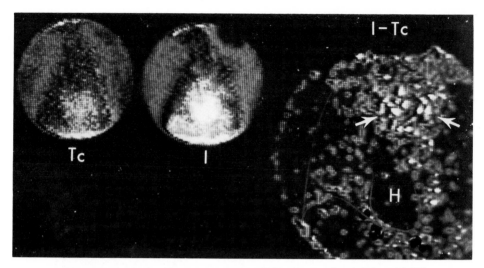

Figure 24. Anterior subtraction image of the chest using I-131 anti-hCG antibody and Tc-99m HSA shows a primary germinal cell tumor (arrows) in the superior mediastinum.

REFERENCES

1. Cutler SJ, Young JL, Jr.: Third national cancer survey: Incidence data. Natl Cancer Inst Mongor 1975; 41:10–27, 100–135, 388–427
2. Say CC, Hari JM: Transitional cell carcinoma of the renal pelvis: Experience from 1940–1972 and literature review. J Urol 1974; 112:438–442
3. Aron BS: Wilm's tumor—A clinical study of 81 patients. Cancer 1974; 33:637–646
4. Silverberg E: Cancer statistics, 1982. Ca-A Cancer J Clinic 1982; 32:15–31
5. Giehring GG, Rodriguez FR, Woodhead DM: Malignant degeneration of cryptorchid testes following rociopexy. J Urol 1974; 112:354–356
6. Weigensberg IJ: Metastatic renal carcinoma: Unusual and deceptive presenting features. South Med J 1972; 65:611–616
7. Latham HS, Kay S: Malignant tumors of the renal pelvis. Surg Gynecol Obstet 1974; 138:613–622
8. Rubin P: Cancer of the urogenital tract: Wilm's tumor and neuroblastoma. JAMA 1968; 204:981–982
9. Bessette PL, Abel MR, Herwig KR: A clinicopathologic study of squamous cell carcinoma of the bladder. J Urol 1974; 112:66–67
10. Jewett HJ: Cancer of the bladder; Diagnosis and staging. Cancer 1973; 32:1072–1074
11. Franks LM: Etiology, epidemiology, and pathology of prostatic cancer. Cancer 1973; 32:1092–1095
12. Borski AA: Diagnosis, staging and natural history of testicular tumors. Cancer 1973; 32:1202–1205
13. Culp DA, Boatman DL, Wilson VB: Testicular tumors; 40 years' experience. J Urol 1973; 110:548–553
14. Ray B, Hajdu SI, Whitmore WF: Distribution of retroperitoneal lymph node metastases in testicular germinal tumors. Cancer 1974; 33:340–348
15. Mostofi FK: Testicular tumors; Epidemiologic, etiologic and patho   · features. Cancer 1973; 32:1186–1201
16. Older RA, Korobkin M, Workman J, et al: Accuracy of radionu   ıaging in

distinguishing renal masses from normal variants. Radiol 1980; 136:443–448

17. Sty JR, Babitt DP, Oechler HW: Evaluating the multicystic kidney. Clin Nucl Med 1980; 5:457–461

18. Kawamura J: Renal cortical imaging and the detection of renal mass lesions. J Nucl Med 1980; 21:494–495

19. Sty JR, Oechler H: Tc-99m glucoheptonate renal imaging: Congenital mesoblastic nephroma. J Nucl Med 1980; 21:809–810

20. Adler J, Greweldinger J, Conradi H: Gallium-67 scans in renal tumors. Urol Radiol 1981; 3:27–29

21. Frankel RS, Richman SD, Gielrud LG, et al: Renal localization of gallium-67 citrate. J Nucl Med 1974; 15:491–494

22. Kawamura J, Itoh H, Yoshida O, et al: "Hot spot" on Ga-67 citrate scan in a case of renal cell carcinoma. Clin Nucl Med 1980; 5:471–472

23. Paterson AHG, Peckham MJ, McCready VR: Value of gallium scanning in seminoma of the testis. Br Med J 1976; 1:1118–1121

24. Johnston GS: Prostate scintiscanning for cancer detection, in Nieburgs HE (ed): Proc 3rd Internat Symp Detection Prev Cancer. New York, Marcel Dekker, 1980, pp 2301–2304.

25. Fairweather DS, Bradwell AR, Dykes PW: Localization of malignant teratoma deposits by emission scanning after injection of radiolabeled antibody to alpha-fetoprotein. Clin Sci 1981; 61:1

26. Pollen JJ, Gerber K, Ashburn WL, et al: The value of nuclear bone imaging in advanced prostatic cancer. J Urol 1981; 125:222–223

27. Oliveux A, Girob JC, Moyses B, et al: Recent advances in isotopic exploration of the skeletal system in prostatic cancer. Ann Urol 1979; 13:167–171

28. Constable AR, Cranage RW: Recognition of the superscan in prostatic bone scintigraphy. Br J Radiol 1981; 54:122–125

29. Stone AR, Merrick MV, Chisholm GD: The bone scan as a monitor of prostatic cancer. Clin Oncol 1980; 6:349–360

30. Anta MA, Remblish R, Wasserman I, et al: Serial bone scan for prostatic bony metastasis during diethylstilbesterol therapy. Invest Radiol 1978; 13:389–391

31. Appel RG, Brandeis WE, Georgi P, et al: Radiographical and scintigraphic appearance of bone metastasizing Wilm's tumor. Ann Radiol 1982; 25:14–18

32. McLorie GA, Orovan WL, Lee JL: Liver scanning in malignant genitourinary tumors. Can J Surg 1980; 23:90–92

33. Belville WD, McLeod DG, Prall RH, et al: The liver scan in urologic oncology. J Urol 1980; 123:901–903.

34. Bailey TB, Pinsky SM, Mittemeyer BT, et al: A new adjuvant in testis tumor imaging: Ga-67 citrate. J Urol 1973; 110:387–392

35. Gardiner RA, Fitzpatrick JM, Constable AR, et al: Improved techniques in radionuclide imaging of prostatic lymph nodes. Br J Urol 1979; 51:561–564

36. Ege GN: Augumented iliopelvic lymphoscintigraphy: Application in the management of genitourinary malignancy. J Urol 1982; 127:265–269

37. Kim EE, Deland FH, Nelson MO, et al: Radioimmunodetection of cancer with radiolabeled antibodies to α-fetoprotein. Cancer Res 1980; 40:3008–3012

38. Goldenberg DM, Kim EE, Deland FH, et al: Clinical radioimmunodetection of cancer with radioactive antibodies to human chorionic gonadotropin. Science 1980; 208:1284–1286

39. Moser E, Marx FJ, Gebauer A, et al: Radionuclide angiography with Tc-99m DTPA to evaluate palliative embolization of renal cell carcinoma. ROEFO 1981; 135:274

40. Rul   D, Coakley AJ: Early detection of lung toxicity after bleomycin therapy. Ca: at Rep 1980; 64:732–734

CHAPTER 8

Gynecologic Neoplasms

INTRODUCTION

Radiologic procedures available to define the site, nature, and extent of gynecologic neoplasms include plain radiography of the abdomen, radiography of the chest, intravenous radiography, gastrointestinal barium study, lymphangiography, angiography, pelvic pneumography, hysterosalpingography, radionuclide scintigraphy, ultrasonography, computed tomography, and nuclear magnetic resonance imaging.

The diagnostic sequence utilized is individualized depending upon the clinical presentation, and is intended to minimize superfluous studies that result in financial burden, inconvenience, radiation exposure, morbidity, and mortality. The availability of equipment and personnel, and expertise in performance and interpretation of the various procedures, must also be considered.

Because of the relatively late stage in which patients with gynecologic tumors often present, the diagnostic effort is directed primarily to establish the extent of the disease.

Radionuclide liver-spleen, bone and brain imagings have been utilized to evaluate metastatic lesions in advanced gynecologic cancer patients. Although computed tomography has higher diagnostic accuracy, a practical screening plan for metastatic lesions consists of scintigraphy, followed only as clinically indicated by CT. Sonography often is used as an alternative to confirm the isotopic findings and to resolve other problems. The nature of the primary gynecologic neoplasms can often be defined by CT or ultrasound. With the advent of CT and ultrasound, angiography is rarely utilized to demonstrate an ovarian mass. The role of NMR imaging is yet to be determined.

In advanced lesions, postirradiation, or post-surgical patients, CT appears useful in the evaluation of lymphatic involvement or obstruction, although it shows much less overall accuracy. Radionuclide lymphangiogram also may be useful in screening lymphatic obstruction.

CHARACTERISTICS OF GYNECOLOGIC CANCER

Epidemiology

In the United States, gynecologic malignancies comprise 28 percent of all cancers in women (endometrium 13 percent, ovary 6 percent, uterine cervix 6 percent, and others 3 percent).

Ovarian cancer is diagnosed in approximately 17,000 women in the United States each year, and is responsible for 11,000 deaths per year, the leading cause of gynecologic cancer death. It has a 5-year survival rate of 29 to 32 percent.[1,2] There is about one carcinoma for every four or five ovarian tumors. In general, serous and mucinous carcinomas are uncommon in the first two decades of life, but increase in frequency thereafter with age. Carcinomas of germ cell origin predominate in children and in young women.[3] Granulosa cell tumors represent about 20 percent of all malignant tumors in young girls.[4]

Endometrial carcinoma accounts for 7.6 percent of all cancer in white women, but only 4.5 percent of cancer in black women in the United States.[5] The overwhelming majority of endometrial carcinomas develop after menopause, and it is frequently found in nulliparous women. Sarcomas make up only about 4 percent of all malignant uterine tumors, and occur in relatively younger women than do the carcinomas. Also, they are seen relatively more frequently in black women.[6]

Carcinoma of the cervix follows carcinoma of the breast, colorectum, and endometrium in incidence of malignant neoplasms in women. In 1982 it made up 5.4 and 13.9 percent of all cancer in white and black women in the United States, respectively.[5] It occurs most often between 45 and 55 years of age. The low occurrence of cervical cancer in Jewish women has long been recognized. Cancer of the cervix is also infrequent in nulliparous women, in those inactive sexually, and in unmarried or childless women.

Cancer of the vagina is rare and comprises 1 to 2 percent of all gynecologic cancers.[7] It is rare among blacks and Jews in the United States. It is usually found in women between 45 and 65 years of age.

Vulvar carcinoma accounts for 0.7 percent of all female cancers, and over two-thirds are found after the age of 60 years.[8]

Pathology and Metastasis

Epithelial or Müllerian cancers of the ovary account for 88 percent of malignant ovarian neoplasms in the adult.[9] Carcinoma accounts for 40 percent; the ratio of benign to malignant epithelial tumors is approximately 1:9. They may be bilateral. Mucinous cystic tumors are less commonly malignant (12 percent), the benign to malignant ratio being 7:1. Rupture of either a benign or malignant tumor may cause pseudomyxoma peritonei. Endometrioma accounts for 15 percent of ovarian tumors and may be associated with carcinoma of the endometrium. Undifferentiated or unclassified adenocarcinoma (15 percent) and clear cell adenocarcinoma (6 percent) complete the list of malignant epithelial neoplasms of the ovary.

The germ cell tumors include the very common benign cystic teratoma or dermoid cyst. They are bilateral in 10 to 15 percent; 1 to 2 percent become malignant. The dysgerminoma is the analogue of the male seminoma. Tumors that contain choriocarcinoma and endodermal sinus tumors are highly malignant. Malignant tumors arising from the ovarian stroma include the feminizing granulosa-theca cell tumor and the virilizing Sertoli-Leydig cell tumor. Only about 10 percent of them are malignant.

Connective tissue tumors comprise 10 percent of ovarian neoplasms. Fibroma, although benign, may be associated with ascites and hydrothorax (Meig's syndrome).

Metastases to the ovaries most frequently originate from neoplasms of the gastrointestinal tract, breast, lung, and reticuloendothelial system. Metastases are usually bilateral. Krukenberg tumor, originally described as bilateral solid ovarian tumors metastatic from a gastric carcinoma, now encompasses all ovarian metastases arising from carcinoma of the gastrointestinal tract.[10]

Cystadenocarcinomas usually spread through the walls of the cyst and implant widely on the peritoneal surface, giving rise to pseudomyxoma peritonei. Para-aortic lymph nodes frequently are involved (78 percent) and also mediastinal lymph nodes (50 percent), inguinal nodes (43 percent), and supraclavicular nodes (26 percent). Para-aortic lymph nodes often are involved in the early stage.[11] Lung metastases are observed in fewer than 10 percent of the cases. Metastases to the liver are even less frequent, and bony metastases are rare. Carcinomas arising from cystic teratomas or sarcomas may metastasize widely. Dysgerminomas may present bony metastases.

Endometrial carcinomas are usually well circumscribed and grow toward the endometrial cavity. They may be preceded by endometrial polyps. They are classified into three major groups: adenocarcinoma, adenoacanthomas, and adenosquamous carcinoma. The majority of pelvic sarcomas in women arise in the uterus as leiomyosarcoma, endometrial stromal sarcoma, or mixed mesodermal sarcoma. Leiomyomas, the most common tumor of the female genital tract, are usually multiple and rarely undergo sarcomatous change.

Lymph node involvement in endometrial adenocarcinoma usually does not appear until the disease is moderately advanced. It occurs in the external iliac chains (about 15 percent) and also in the para-aortic and inguinal regions. Retrograde permeation of lymphatics results in implants in the lower third of the vagina and vulva. Distant metastases to the liver, lungs, brain, and skeleton are observed mostly in advanced patients. Sarcomas metastasize frequently to the lungs (in about 80 percent of the cases) and also to the lymph nodes (about 50 percent of the cases).[12]

In one series of 259 carcinomas of the cervix,[13] 95 percent were squamous cell carcinomas, 3 percent were adenocarcinomas, 1 percent were mesonephric carcinomas, and 0.8 percent were adenosquamous carcinomas. Stromal sarcoma and carcinoma of the endometrium or vagina occasionally involve the cervix.

Carcinomas of the cervix usually spread to the lymph nodes of the external iliac and hypogastric groups. Extension to the lumboaortic chain of nodes, mediastinum, and supraclavicular lymph nodes occurs by continued permeation of lymphatic channels. Hematogenous spread may take place though the portal and caval venous systems and their anastomoses with the vaginal vein plexus. Metastases to the liver, lungs, brain, and bones occur infrequently. In a series of 2,200 patients treated for carcinoma of the cervix, 15 percent manifested distant metastases (most commonly in the lung); the proportion of distant metastases increased with the stage of the treated tumors.[14]

Carcinoma of the vagina most often develops on the upper third of the vagina. The overwhelming majority of tumors of the vagina are squamous cell carcinomas. Vaginal sarcomas occurring in young girls are of the botryoid type and arise most often on the anterior wall. The sarcoma occurring in adults may arise from any part of the wall and is most often parietal rather than mucous origin.

Carcinomas of the vagina metastasize to the lymph nodes of the external iliac and hypogastric chains, with inguinal node involvement observed only when the vulva has been invaded. Distant visceral metastases are rare except with sarcomas in the adult.

The greatest majority of carcinomas of the vulva develop on the labia, and they are squamous and well differentiated in nature.

A high proportion of patients with vulvar carcinoma develop metastases to the inguinal lymph nodes. Metastases are often bilateral. Metastasis to pelvic lymph nodes has been found in 7 to 16 percent of the patients.[15]

Clinical Evolution

In general, ovarian tumors grow insidiously and can reach a huge size before causing enough symptoms to bring the patient to the physician. Pelvic discomfort or pain (57 percent), abdominal distention (51 percent), and bleeding (25 percent) are usually manifestations of abdominal disease. After the masses become very bulky, invasion of the bladder or large bowel may occur. These changes result in infection or intestinal obstruction. The serous cystadenocarcinoma and the undifferentiated carcinomas often metastasize distantly, particularly to lymph nodes, the evolution being quite rapid due to widespread dissemination.

Most carcinomas of the endometrium develop slowly, and even though the disease may be considerably advanced, the patients may survive for years in rather good general condition. The most common early sign of the disease is a slight vaginal bleeding. Watery discharge is present at times, and is a significant sign of the disease.

The symptoms that could betray the existence of an early carcinoma of the cervix are easily taken for inconsequential irregularities of menstruation. A yellow vaginal discharge is present in the majority of patients with advanced disease. Pain and weight loss almost invariably present with carcinoma of the cervix.

The clinical onset of carcinoma of the vagina is often manifested by vaginal bleeding of variable intensity. The second most frequent complaint is vaginal discharge.

Most cases of vulva carcinoma are preceded or accompanied by pruritus, often due to kraurosis. There is, sometimes, intermittent bleeding and also

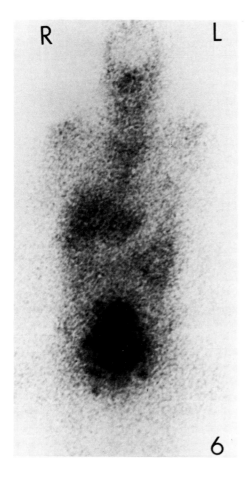

Figure 1. Longitudinal tomographic image of whole body 48 hours after Ga-67 citrate injection shows marked accumulation of radiogallium in the mid-pelvis with uterine lymphoma and bilateral inguinal lymph nodes.

dyspareunia. Generally, carcinomas of the vulva are elevated outgrowths that may become very extensive with associated cellulitis and phlebitis. Involvement of the inguinal nodes occurs rather early in the development of the disease.

NUCLEAR IMAGING AND OTHER PROCEDURES

Imaging of Primary Tumors

Kida et al.[16] evaluated In-111 bleomycin scintigraphy for malignant tumors of the female genital tract. Of 22 patients with uterine cervix tumors, four patients with ovarian tumors, one patient with endometrial tumor, one patient with vulvar tumor, and one patient with chorioepithelioma, 23 patients (79 percent) had positive images, and the technique also was valuable in assessing the extent of parametrial invasion. Compared with ultrasonography, In-111 bleomycin scans were found to be complementary in ten patients with ovarian cancer and five patients with endometrial cancer.[17] In no patient were In-111 and ultrasound studies both misleading.

   Ga-67 citrate imaging (Figs. 1 and 2) in 32 patients with gynecologic

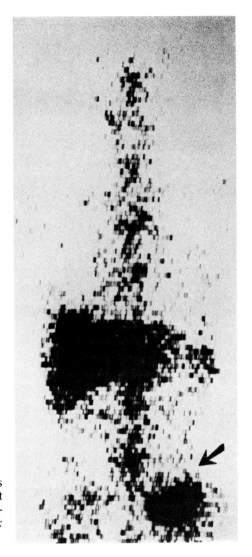

Figure 2. Ga-67 whole-body image shows heavy uptake of radiogallium in the left upper pelvic area (arrow) due to an ovarian carcinoma. *(From Gynecol Oncol 1978; 6:134, with permission.)*

malignant tumors successfully established the presence of tumors in 53.2 percent and their absence in 21 percent.[18] Forty-eight women with pelvic masses were evaluated by Tc-99m phosphate and gray scale echographic scans.[19] Twenty-one of 24 who displayed results consistent with diagnosis of leiomyoma uteri were confirmed pathologically, and 29 of 30 who had scintigraphic patterns considered characteristic of ovarian cyst had this diagnosis at surgery. There has been a report on positive imaging of arrhenoblastoma of the ovary with I-131 aldosterol.[20]

All primary ovarian cancers in 13 patients could be localized with I-131-labeled antibodies to carcinoembryonic antigen.[21]

Imaging of Metastatic Tumors
One-hundred and one patients with gynecologic cancers received radionuclide liver imagings as a part of their staging workup. Only 1 of 68 patients

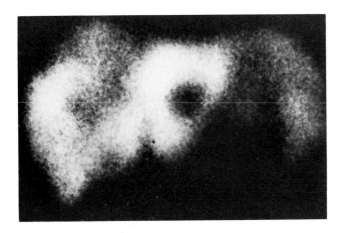

Figure 3. Anterior image of liver-spleen with Tc-99m sulfur colloid shows multiple metastatic lesions of ovarian carcinoma in right and left lobes of the liver.

with stage I-III disease and 8 of 33 patients with stage IV disease had positive liver scan (Figs. 3 and 4) results consistent with hepatic metastases.[22]

Staging radionuclide bone imagings or radiographic bone surveys were obtained in 97 patients with endometrial carcinoma. Of the 77 patients with stage I or II disease, no metastasis was identified. Three patients demonstrated bony metastases that were detectable by bone scans and bone surveys.[23] They all conclude that liver and bone scanning in asymptomatic patients with early stage gynecologic cancers may not be warranted as staging procedures.

Bone scans in 14 patients with cervical carcinoma (Fig. 5) showed renal asymmetry; 11 of these had obstructive uropathy due to tumor invasion or

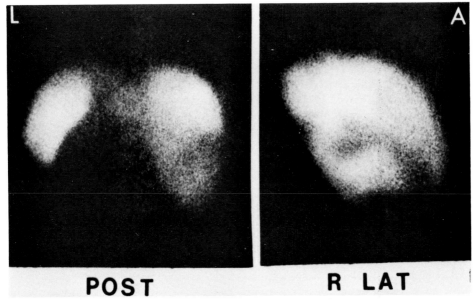

Figure 4. Posterior and right lateral views of liver-spleen with Tc-99m sulfur colloid show metastatic lesions of cervix carcinoma in the right lobe of the liver.

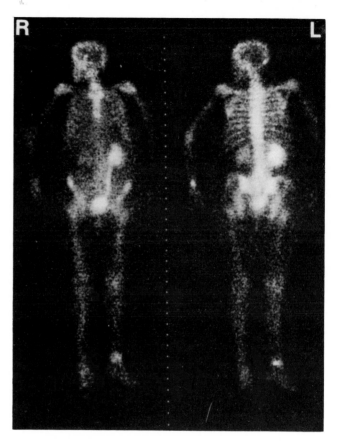

Figure 5. Whole-body images with Tc-99m MDP show obstructive uropathy on the left due to cervical carcinoma.

radiation therapy.[24] Bone scans may also be of value in staging and follow-up of ovarian carcinoma (Fig. 6).

Metastatic serosal and omental implants from a primary papillary endometrial carcinoma have been demonstrated on Tc-99m methylene diphosphonate bone scans.[25] Intravenous urograms showed calcifications in the pelvic masses as well as abdominal soft tissues.

Ga-67 uptake was demonstrated in the metastasis of cervical cancer in the left psoas bed.[26]

Brain metastasis occurs in 10 percent of choriocarcinoma, and radionuclide brain imaging may be useful in patients with neurologic symptoms or signs to rule out a metastatic cerebral lesion.

Metastatic choriocarcinoma was localized by I-131 labeled to antibodies of beta human chorionic gonadotropin (Fig. 7) and resected allowing cure in one patient after multiple chemotherapies failed.[27] Metastatic lesions of ovarian cancer in six of nine cases were also detected with I-131 antibodies to CEA.[21]

Lymphoscintigraphy of the pelvis has been described using intralymphatic injection of Tc-99m colloids or Ga-67 salts and subcutaneous injection of Tc-99m antimony colloid,[28] and demonstrated a limited role in evaluating metastatic lesions. In three patients with squamous cell carcinoma of the

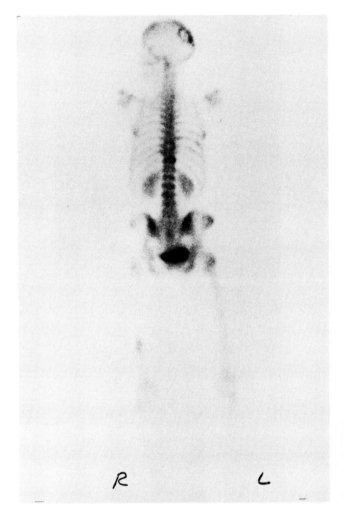

R   L

Figure 6. Posterior image of skeleton with Tc-99m MDP shows metastatic lesions of ovarian carcinoma in the skull, chest, spine, pelvis, and femora.

vulva, sequestration of radioactivity was detected in inguinal regions following bilateral foot injection of I-131 CEA antibody.[29] Abdominal-mediastinal lymphoscintigraphy was obtained using intraperitoneal injection of Tc-99m sulfur colloid in 21 patients with ovarian carcinoma and normal diaphragm, and 19 of them (90 percent) showed normal visualization of thoracic nodes.[30] There was a poor relationship between node visualization and prognosis.

Of 154 Chinese patients who underwent gynecological operations, four showed a positive I-125 fibrinogen leg scan for venous thrombosis; in those who had Wertheim hysterectomy for carcinoma of the cervix, the incidence of abnormal fibrinogen scan was 6.7 percent.[31]

Tc-99m macroaggregated albumin angiography and perfusion studies (Fig. 8) have been very helpful for intra-arterial infusion chemotherapy of gynecologic cancers by recognizing displaced or misplaced catheters and estimating approximate division of chemotherapeutic agents in bilateral pelvic catheters.[32]

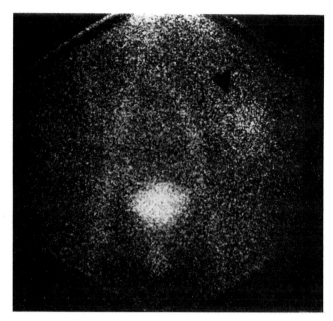

Figure 7. Anterior pelvic image 48 hours after injection of I-131 hCG antibodies shows a focal uptake of radioactivity in left mid-quadrant with choriocarcinoma. (*From Gynecol Oncol 1980; 10:257, with permission.*)

OTHER RADIOLOGIC PROCEDURES

Chest radiography is useful for detecting pulmonary metastases and pleural effusions. Routine chest tomography is not recommended since lung metastasis usually precedes other findings such as ascites.

Plain radiography of the abdomen may reveal a nonspecific mass displacing bowel loops. Calcification may indicate uterine fibroid and ovarian cystadenoma or cystadenocarcinoma.

Intravenous urography may show a pelvic mass distorting the bladder or obstructing the ureters.

Barium enema has been used to define extension from the pelvic neoplasms or its metastases. It occasionally detects invasion of the bowel.

The accuracy of ultrasound in the detection of a pelvic mass and in determining its size, location, and consistency is as high as 91 percent,[33] although it is the most technically demanding procedure. However, the abnormality is nonspecific except for a few conditions with specific features that suggest the histologic diagnosis. Because of the absence of ionizing radiation, sonography is useful in defining an ovarian mass in the pregnant patient.

With the ability of computed tomography to increase resolution and to demonstrate subtle differences by detecting fat, calcification, cystic or solid mass, or septation. Local and extra-pelvic extension of the tumor is recognized by the invasion of adjacent viscera, or retroperitoneal lymph nodes. CT is the best method for the detection of liver or brain metastases, but it practically follows screening radionuclide imaging with clinical indications.

Contrast lymphangiography is less utilized because of difficulties of the technique and intrepretation. Positive lymphangiograms have been found in

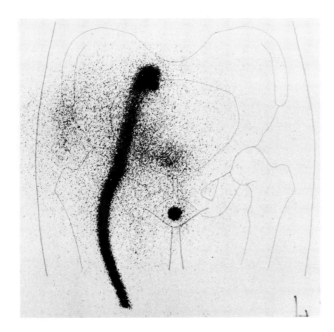

Figure 8. Anterior pelvic image after infusion of Tc-99m MAA into intra-arterial catheter (outlined by curvilinear radioactivity) shows activities distributed in right true pelvis and buttock areas, indicating the catheter tip placed in right internal iliac artery and slightly more perfusion through anterior division. Hot spot denotes a marker on symphysis pubis.

21 percent of 289 patients with stage I or stage II ovarian carcinoma, and in 19 percent of 76 patients with carcinoma of the uterine corpus.[34,35]

Angiography is reserved as the last resort to solve a clinical problem. Most ovarian malignancies and their metastases are relatively hypovascular while uterine leiomyoma and trophoblastic neoplasms are usually hypervascular. Small metastases in the liver or brain can be opacified by superselective angiography.

In recent years, diagnostic radiologists, guided by fluoroscopy, ultrasonography, and CT, have employed percutaneous needle biopsy to establish the diagnosis of lesions throughout the body, including pelvic masses.

REFERENCES

1. American Cancer Society Publications: Cancer, Facts and Figures, 1980. New York, 1980
2. Cutler SJ, Myers MH, Green SB: Trends in survival rates of patients with cancer. N Engl J Med 1975; 293:122–124
3. Lingeman CH: Etiology of cancer of the human ovary: A review. J Natl Cancer Inst 1974; 53:1603–1618
4. Li FP, Fraumeni JF, Dalager N: Ovarian cancers in the young. Cancer 1973; 32:969–972
5. Culter SJ, Young JL: Third national cancer survey: Incidence data. Natl Cancer Inst Monogr 1975; 41:10–27, 100–135, 388–427
6. Christopherson WM, Williamson ED, Gray LA: Leiomyosarcoma of the uterus. Cancer 1972; 29:1512–1517
7. Perez CA, Arneson AN, Galakatos A, et al: Malignant tumors of the vagina. Cancer 1973; 31:36–44

8. Frankendal B, Larson LG, Westhing P: Carcinoma of the vulva. Acta Radiol 1967; 12:165–174

9. Rutledge FN, Fletcher GH, Smith JP, et al: Gynecologic cancer, in Clark RL, Howe CH, (eds): Cancer Patient Care. Chicago, Year Book Medical Publishers, 1976, pp 263–308

10. Sanders RC, James AE: The Principles and Practice of Ultrasonography in Obstetrics and Gynecology, ed 2. New York, Appleton-Century-Crofts, 1980

11. Knapp RC, Friedman EA: Aortic lymph node metastases in early ovarian cancer. Am J Obstet Gynecol 1974; 119:1013–1017

12. Taylor HB, Norris HJ: Mesenchymal tumors of the uterus. Arch Pathol 1966; 82:40–44

13. Stage AH, Crawford EJ, Robinson LS, et al: Combined radiologic/operative therapy in the treatment of cervical malignancy. Am J Obstet Gyncol 1974; 120:960–968

14. Carlson V, Delclos L, Fletcher GH: Distant metastases in squamous cell carcinoma of the uterine cervix. Radiol 1967; 88:961–966

15. DiSaia PJ, Morrow CP, Townsend DE: Cancer of the vulva. Calif Med 1973; 118:13–18

16. Kida T, Ikeda M, Saito M: Diagnostic value of In-111 bleomycin scintigraphy for malignant tumors of the female genital tract. Radioisotopes 1978; 27:514–519

17. Alberts DS, Woolfenden JM, Habert K, et al: Comparison of ultrasonic and In-111 bleomycin scanning of gynecologic tumors. Gynecol Oncol 1978; 6:145–151

18. Bloomfield RD, Sy WM, Yoon T, et al: The use of Ga-67 citrate in gynecologic malignancies. Gynecol Oncol 1978; 6:130–137

19. Manfredi OL, Aruny JE: The role of Tc-99m bound phosphates and grey scale echography in the differentiation of pelvic tumors. Clin Nucl Med 1979; 4:99–107

20. Nakajo M, Sakata M, Shinohara S: Positive imaging of arrhenoblastoma of the ovary with I-131 adosterol. Radioisotopes 1978; 27:407–409

21. Va Nagell Jr, Kim EE, Casper S, et al: Radioimmunodetection of primary and metastatic ovarian cancer using radiolabeled antibodies to carcinoembryonic antigen. Cancer Res 1980; 40:502–506

22. Harbert JC, Rocha L, Smith FP et al: The efficacy of radionuclide liver and bone scans in the evaluation of gynecologic cancers. Cancer 1982: 49:1040–1042

23. Mattler FA, Christie JA, Garcia JF, et al: Radionuclide liver and bone scanning in the evaluation of patients with endometrial carcinoma. Radiol 1981; 141:777–780

24. Katz Rd, Alderson PO, Rosenshein NB, et al: Utility of bone scanning in detecting occult skeletal metastasis from cervical carcinoma. Radiol 1979; 133:469–472

25. Cannon JR, Long RF, Berens SV, et al: Metastatic abdominal implants of endometrial carcinoma demonstrated on Tc-99m methylene diphosphonate bone scan. Clin Nucl Med 1978; 3:310–311

26. Smith RW, Geoneratne NS: Psoas bed gallium uptake in a patient with carcinoma of the cervix. Clin Nucl Med 1978; 3:230

27. Hatch KD, Mann WJ, Boots LR, et al: Localization of choriocarcinoma by I-131 beta hCG antibodies. Gynecol Oncol 1980; 10:253–261

28. Merrick MV, Kirkpatric AE: Lymphoscintigraphy of the pelvis. Br J Radiol 1979; 52:337–339

29. Deland FH, Kim EE, Goldenberg DM: Noninvasive lymphoscintigraphy and tumor-associated antigens. Compre Therapy 1980; 6:68–72

30. Katz RD, Rosenshein NB, Alderson PO, et al: Abdominal-mediastinal lymphoscintigraphy in patients with ovarian carcinoma. J Nucl Med 1979; 20:646–649

31. Tso SC, Wong V, Chan V, et al: Deep vein thrombosis and changes in coagulation and fibrinolysis after gynecological operations in Chinese. Br J Haematol 1980; 46:603–612

32. Kim EE, Bledin AG, Kavanagh J, et al: Chemotherapy of cervical carcinoma: Use of Tc-99m MAA infusion to predict drug distribution. Radiol 1984; 150:677–681

33. Lawson TL, Albarelli JN: Diagnosis of gynecologic pelvic masses by gray scale ultrasonography: Analysis of specificity and accuracy. Am J Roentgenol 1977; 128:1003–1006

34. Fuks Z: External radiotherapy of ovarian cancer: Standard approaches and new frontiers. Semin Oncol 1975; 2:253–266

35. Douglas B, MacDonald JS, Baker JW: Lymphography in carcinoma of the uterus. Clin Radiol 1972; 23:286–294

CHAPTER 9

Musculoskeletal Neoplasms

BONE TUMORS

During the past decade radionuclide bone imaging has become one of the most important investigations available in nuclear medicine. Many physicians rely on the bone scan to provide highly significant clinical information about bony abnormalities. Many clinical studies have compared the relative efficiency of the radionuclide bone scan and radiographic bone survey in the detection of bone metastases. These studies agree that the bone scan is a more sensitive indicator of the presence of bone metastases than radiography. In contrast to the radiograph, which demonstrates the net result of bone destruction and repair, the bone scan is based on the dynamic response of bone to the disease process. Tumor destruction of bone results in a local increase in bone-blood flow and osteoblastic activity due to a reactive formation of new bone. The greater the reactive process, the more intense the abnormality seen on the bone scan. Therefore, the bone scan also has a predominant role in the assessment of response to systemic therapy in patients with primary or metastatic bone lesions. When lesions heal under the influence of treatment, avidity for radiopharmaceuticals ceases with normalization of the bone scan, and when lesions progress, the response in new bone formation results in increased uptake of radioactivity.

PRIMARY BONE TUMORS

Classification and Characteristics
Table 1 gives the classifications of the primary benign and malignant bone tumors, and Table 2 describes key characteristics of malignant bone tumors.

TABLE 1. CLASSIFICATION OF PRIMARY BONE TUMORS

Origin	Benign	Malignant

Bone	Osteoma
	Osteoid osteoma
	Osteoblastoma
		Osteosarcoma
		    Parosteal or periosteal
		    Arising in Paget's disease
		    Irradiation or infarcts

Cartilage	Chondroblastoma
	Chondromyxoid
	    fibroma
	Osteochondroma
	Enchondroma
	Juxtacortical chondroma
		Mesenchymal
		    Dedifferentiated

Marrow		Chondrosarcoma Ewing's sarcoma Plasma cell myeloma Malignant lymphoma

Fibrous connective tissue	Lipoma
	Desmoplastic fibroma
	Periosteal desmoid
	    fibromyxoma
		Liposarcoma
		Fibrosarcoma
		Malignant fibrous
		    histocytoma

Smooth muscle	—	Leiomyosarcoma
Vascular	Hemangioma Glomus tumor Angiomatosis Hemangiopericytoma Lymphangioma	Angiosarcoma Hemangioendothelioma Hemangiopericytoma
Neurogenous tissue	Neurilemoma Neurofibromatosis Ganglioneuroma	—
Notochord	—	Chordoma
Mixed	—	Osteoliposarcoma
Uncertain	Giant cell tumor	Giant cell tumor Adamantinoma

(Modified from Spjut HJ, et al: Tumors of bone and cartilage, in Atlas of Tumor Pathology, sect 2, fasc 5, Washington, D.C., A.F.I.P., 1971.)

Bone Scans in Primary Benign Bone Tumors

Apart from occasional case reports, there appears to be little information in the literature concerning the role of bone scanning in the evaluation of benign bone tumors. Gilday and Ash[1] have described the scan findings of 19 osteoid osteomas, the most common benign bone tumor. The scan in these patients was very useful in locating the cause of bone pain and in evaluating whether or not a radiological lesion was indeed benign and solitary. Multiple views and pinhole magnification were especially valuable in diagnosing lesions in the spine and pelvis, which are not always radiographically apparent.

The bone scan usually demonstrates a focal area of abnormally increased radioactivity in osteoid osteoma, and the abnormality may be very prominent in ivory osteomas of the skull and highly vascular osteoblastomas of the axial skeleton.[2]

The blood-pool images in 42 osteoid osteomas were performed after injection of Tc-99m MDP that showed a small hyperemic lesion in all cases of osteoid osteoma, thus distinguishing this tumor from other lesions that may show uptake of radionuclide activity after 2 hours (Fig. 1).[3]

TABLE 2. CHARACTERISTICS OF COMMON MALIGNANT BONE TUMORS

	Osteosarcoma	Ewing's Sarcoma	Chondrosarcoma	Multiple Myeloma	Giant Cell Tumor
Predominant sex	M	M	M	M	F
Age (years)	10–30	4–20	>40	>40	20–35
Sites	Lower femur Upper tibia Upper humerus	Femur Tibia Mandible	Pelvis Long bones	Pelvis Femur	Lower femur Upper tibia Lower radius
Location	Metaphysis	Shaft	Metaphysis and disphysis	Shaft	Epiphysis and metaphysis
Metastases	Lungs	Lymph nodes, lungs, skull, ribs, spine	Lungs	Skull, ribs, spine	Infrequent

(Modified from Malignant tumors of bone, in Ackerman and del Regato's Cancer, ed 5, St. Louis, Illinois, C.V. Mosby, 1977, chap 19.)

Gilday and Ash[1] also have studied other types of benign tumors by bone scan, including simple and aneurysmal bone cysts, fibrous cortical defects, and nonossifying fibromas, all of which had minimal or no increased radioactivity unless traumatized. Although osteochondromas (Fig. 2) and enchondromas demonstrated varied accumulation of the radioactivity, the bone scan was helpful in differentiating these from sarcomatous lesions by using blood-pool

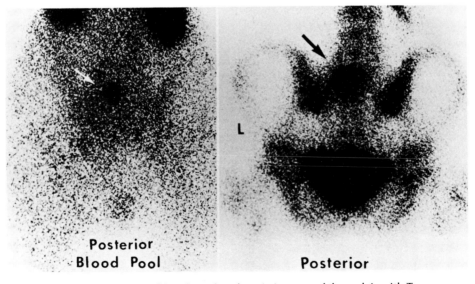

Posterior
Blood Pool

L

Posterior

Figure 1. Posterior blood-pool and static images of the pelvis with Tc-99m MDP show an osteoid osteoma (arrows). *(From Radiol 1983; 137:193, with permission.)*

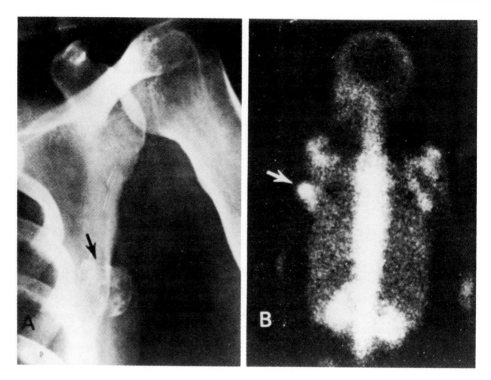

Figure 2. Radiograph (**a**) of left scapula and left shoulder shows an osteochondroma (arrow) of scapula. Posterior view of upper trunk with Tc-99m EHDP (**b**) shows an intense focal activity in scapular osteochondroma (arrow). *(From Am J Roentgenol 1978; 130:332, with permission.)*

images. The absence of hyperemia in the immediate postinjection period favors the diagnosis of a benign tumor, as does low-grade radioactivity on the delayed images.[1]

Humphry et al.,[4] however, reported three cases of chondroblastoma (Figs. 3 and 4) that were hyperemic and avidly accumulated Tc-99m MDP. Three cases of nonosteogenic fibroma or fibrous cortical defect demonstrated minimal to mild increased uptake of Tc-99m MDP and thus helped to distinguish these lesions from other benign or malignant abnormalities.[5] The two types of lesion were not fundamentally different from one another.

The bone scan in the reticuloses (eosinophilic granuloma) has also shown abnormal uptake of radioactivity, and aided in arriving at the prognosis and treatment of histiocytic bone lesions.[1] However, the bone scan proved to be less sensitive than radiographs in children with histiocytes X.[6]

Bone Scans in Primary Malignant Bone Tumors
Osteosarcoma is the most common primary malignant tumor of bone and the most important to consider in the context of bone scanning, since the concentration of bone seeking agents is very high due to large amounts of immature osteoid. The majority of osteosarcomas are found in patients 15 to

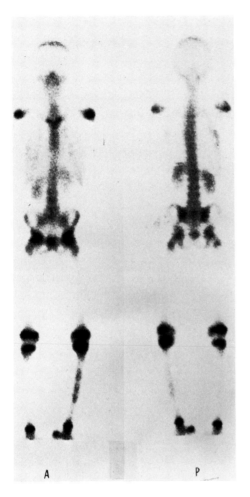

Figure 3. Anterior and posterior whole-body images with Tc-99m EHDP show an increased activity in chondroblastoma of left tibia.

A   P

30 years of age, and they predominate in males. The pain is minimal in most instances, and precedes the appearance of tumor by days to months. Osteosarcomas metastasize primarily to the lungs, and about 5 percent may metastasize to regional lymph nodes.

Metastases to other bones have been well known, and patients with osteosarcoma showed an almost linear increase with approximately 1 percent per month in the occurrence of bone metastases between 5 and 29 months after diagnosis.[7] While pulmonary metastases were always detected prior to bone metastases in the era before adjuvant chemotherapy, in the study of patients with adjuvant chemotherapy 16 percent of patients with metastasis showed osseous metastases prior to or without pulmonary metastases.[8] The tumor on the bone scan appears as a focal area of intense uptake of radioactivity in a characteristic location, usually at the distal end of the femur or proximal end of the tibia (Fig. 5). Goldman and Braunstein[9] also demonstrated diffusely increased radioactivity in large bones with osteosarcoma (Figs. 6 and 7) probably due to regional hyperemia, which may overestimate

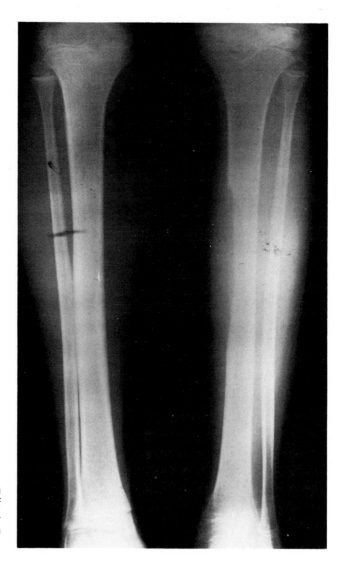

Figure 4. Same tibia as in Fig. 3. Radiographs of lower legs show an osteolytic chondroblastoma in left tibia.

the extent of local involvement. There may be extensive bone destruction in some cases, but increased uptake is also a predominant scan finding in these cases of lytic osteosarcoma. The "skip metastases," which occur frequently in osteosarcoma, have not been demonstrated by the scan using Tc-99m polyphosphate,[10] and the scan has not revealed greater intramedullary extension of tumor than the radiograph.

Because of the osteoid tissue in the tumor, some soft tissue and visceral metastases of osteosarcoma take up bone-seeking radioisotopes, and this has been utilized to improve the early diagnosis of lung metastases.[11] However, plain chest radiographs appear more sensitive for lung metastasis than bone scan.[12] It may be possible to irradiate microscopic metastatic disease with P-32 labeled diphosphonate.[13]

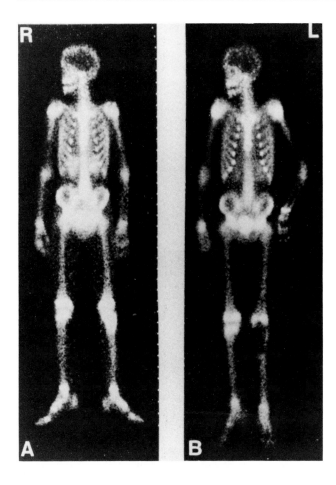

Figure 5. Preoperative (**a**) anterior view of whole-body bone scan with Tc-99m MDP shows an increased activity in osteogenic sarcoma of left proximal tibia. Postoperative (**b**) scan shows a surgical defect.

Ewing's sarcoma (Fig. 8) makes up between 7 and 15 percent of all malignant bone tumors. About two-thirds of them occur in males, and 80 percent are diagnosed in patients under 30 years of age. The first symptom of Ewing's sarcoma is often pain, which usually appears 4 to 5 months before medical attention.[14]

Pathological fractures may take place in Ewing's sarcoma patients (2 to 5 percent) and they have a tendency to heal spontaneously. Ewing's sarcomas metastasize early (nearly 20 percent of patients at initial examination) and widely to lungs, lymph nodes, and other bones, especially the skull. It is still a question whether the bone lesions represent metastases from the primary lesion or multiple foci of origin. Unlike osteosarcoma, Ewing's sarcoma contains no osteoid tissue. Therefore, the uptake of bone-seeking radioisotope is probably due to reactive bone formation and is usually less than in osteosarcoma (Fig. 9). The bone scan provided better definition of the primary tumor than the radiography, and is of more value than either radiography or Ga-67 scanning (Figs. 10 and 11) in the detection of bone metastases in patients with Ewing's sarcoma.[15] In 28 cases of Ewing's sarcoma, bone scans demonstrated bone metastasis in three patients at presentation. In follow-up

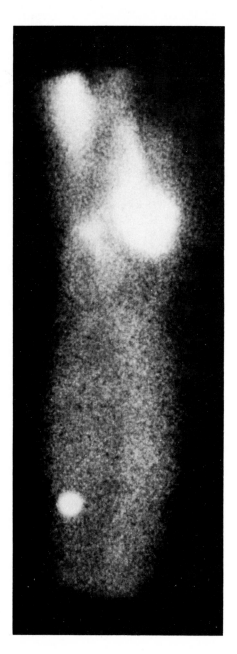

Figure 6. Static image of right knee following Tc-99m MAA infusion through intra-arterial catheter for chemotherapy shows an intense accumulation of activity in the osteogenic sarcoma of right distal femur.

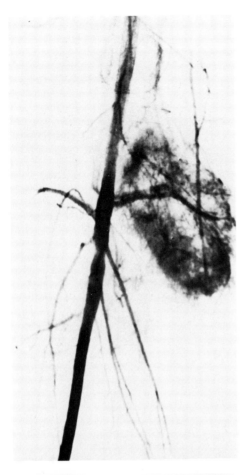

Figure 7. Same femur as in Fig. 6. Contrast angiographic image also shows a vascular blush in osteogenic sarcoma.

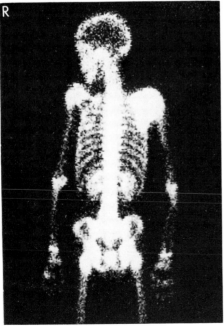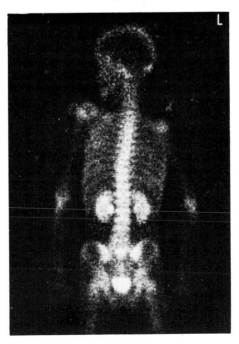

Figure 8. Posterior whole-body image with Tc-99m MDP before the surgery (left) and the chemotherapy using adriamycin shows markedly increased activity in Ewing's sarcoma of the left scapula.

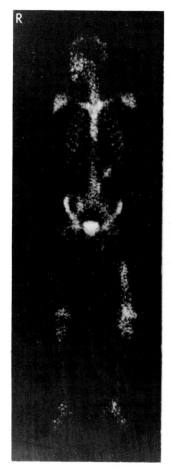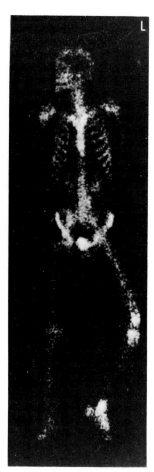

Figure 9. Anterior whole-body image of bone scan using Tc-99m MDP before adriamycin therapy (left) shows an increased activity in Ewing's sarcoma of left femoral shaft, with improvement apparent after therapy (right).

of 22 patients free of metastases at presentation, ten subsequently developed bone metastases. In six of these patients, the bone scan was the earliest demonstrator of metastatic disease.[16] Extraskeletal metastases do not concentrate significant amounts of bone scan agents.

Multiple myeloma occurs predominantly in men past 50 years of age, and the first symptom is usually local pain that often becomes worse with exercise. The tumor (Fig. 12) frequency involves the ribs, vertebrae, pelvis, and flat bones of the skull, and fractures occur in approximately 20 percent of the patients. At autopsy, extraskeletal manifestations are fairly common (65 percent), but in significantly less proportion than the bony manifestations: spleen, liver, and lymph nodes are frequently involved, but also pancreas, thyroid, lungs, and adrenal glands.

Myelomas in bone are usually lytic radiographically, and the radionuclide image (Fig. 13) was relatively insensitive in detecting myeloma at 27 percent of 562 sites.[17] On a limited number of serial images, however, there were seven sites at which a scintigraphic abnormality preceded the radiographic abnormality. It has been suggested that the myeloma cells compete with the osteoblasts for the circulating nutrients necessary for osteogenesis.

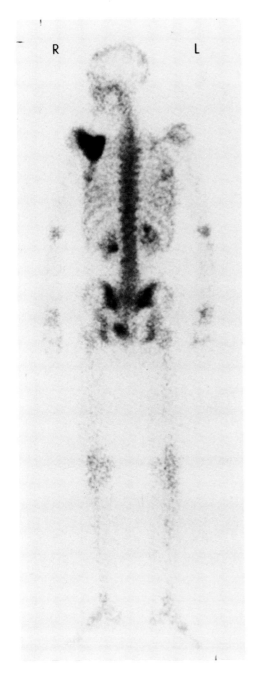

Figure 10. Posterior whole-body bone scan with Tc-99m MDP shows an intense activity in Ewing's sarcoma of right scapula.

Bone Marrow Scans in Bone Tumors

Bone marrow scintigraphy has been relatively underutilized, but is potentially a very important organ-imaging study in the patient with hematologic disease or malignancy. In the patient with malignant disease, it appears practically a simple matter to extend the routine liver-spleen imaging into a complete bone marrow survey, possibly increasing the likelihood of defining the presence of tumor extent. Siddiqui et al.[18] reviewed bone marrow scans

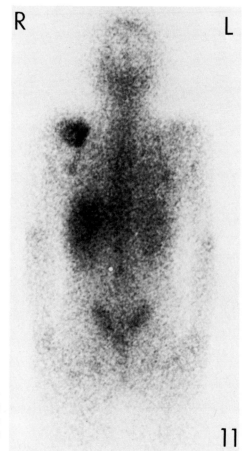

Figure 11. Same image as in Fig. 10. Longitudinal tomographic image of whole body using Ga-67 citrate also shows an increased activity in right scapular Ewing's sarcoma.

using Tc-99m sulfur colloid in 56 pediatric oncology patients. The tumor replacement of the marrow was reflected in the scans, and the extent of the scan defect paralleled the course of the disease. In four patients, despite normal bone scans and radiographs, marrow scan abnormalities due to tumor replacement were present and confirmed by biopsy. In two other patients, the marrow scan abnormality preceded radiographic and histologic evidence of tumor metastasis.

It also has been reported that the In-111 bone marrow scan allows the clinician to avoid selecting a biopsy site with a high risk for sampling error in patients with lymphoma.[19]

Malignant lymphomas primary in bone are rare, and they predominate in males in the second and third decades of life. Local swelling is often a first symptom, but patients are often in good general condition, even with widespread involvement. They metastasize to other bones in about 50 percent of cases and also to regional lymph nodes, liver, and lungs.

A wide variety of bone marrow scan patterns were found in lymphomas.[20] The most common abnormality was bone marrow hypoplasia, which was seen in 45 percent of the cases. Focal defects in the marrow were seen in 20 percent of the patients.

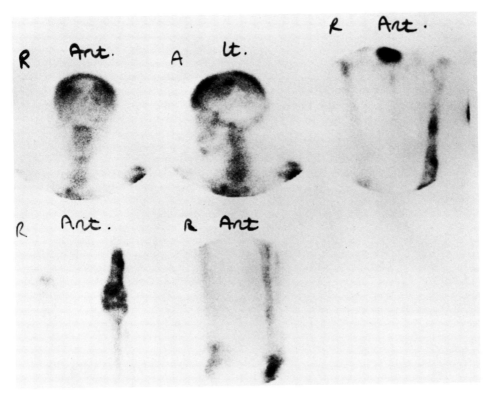

Figure 12. Static images of skull, femurs, and tibias with Tc-99m HEDP show multiple focal areas of increased activity in the skull, left femur, and tibia from involvement by multiple myeloma.

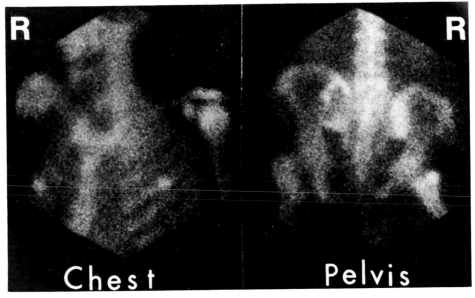

Figure 13. Left anterior oblique view of chest and posterior view of pelvis with Tc-99m MDP show focal increased activities in left and right fifth ribs, and focal decreased activity in left posterior iliac crest. Biopsy confirmed an involvement of multiple myeloma.

Abnormal bone marrow scans were found in all 12 patients with multiple myeloma.[20] Ten patients showed marrow hypoplasia; focal defects were seen in two patients and peripheral expansion in five patients. In the patients who required a biopsy, a lack of marrow uptake in the scan correlated with an almost complete replacement of the marrow with plasma cells. Depending on stage, partial to extensive ablation of axial marrow with compensatory activity around knees was observed, and the scan also aided in understanding of the objective response to marrow ablation.[21]

SOFT TISSUE SARCOMAS

Characteristics

Soft tissue sarcomas constitute less than 1 percent of all malignant tumors,[22] and they are a group of tumors of mesodermal origin (smooth and striated muscle, fat, connective tissue, cartilage, and the vasculature) that appear at any site where the parent tissue is present.

In the retroperitoneal space, liposarcomas are the most frequent primary lesion, followed by the leiomyosarcomas. Fibrosarcomas are the most frequently found tumors in the extremities. Rhabdomyosarcomas are predominant among tumors found in the orbital, gluteal, and interscapular regions. Synovial sarcomas are the most frequent tumors occurring in the region of the ankles and knees. Soft tissue sarcomas are discovered either as a palpable or visible mass by the patient or because of symptoms caused by encroachment upon vital organs or nerves. Most soft tissue sarcomas metastasize primarily to the lungs, to the liver, and to various other organs. Metastases to regional lymph nodes are uncommon in fibrosarcomas, but may be encountered in about half of the patients with rhabdomyosarcomas and synovial, clear cell, and epithelial sarcomas.

Gallium Scan in Soft Tissue Sarcoma

There appears to be less data available regarding the role of gallium imaging in patients with soft tissue sarcomas. A report by Kaufman et al.[23] included gallium scans of 27 soft tissue sarcomas, of which 25 were positive. This included ten tumors that arose in the extremities and limb girdles. Nine of the latter scans were positive. Bitran et al.[24] also reported a series of gallium scans that included 29 soft tissue sarcomas. The accuracy of the scans varied with the specific diagnosis. In one group of 16 tumors including rhabdomyosarcoma, 29 of 31 sites of tumor were positive (94 percent) on the scan. However, only 5 of 16 sites (31 percent) were positive in another group of 13 tumors including liposarcoma, leiomyosarcoma, malignant fibrous histiocytoma, and synovioma.

Pinsky and Henkin[25] discussed the use of Ga-67 citrate in scanning soft tissue tumors and concluded that gallium is still the best available for tumor scanning for various epithelial and nonepithelial tumors.

Other Scannings in Soft Tissue Sarcomas

There have been reports of positive imaging of a liposarcoma using Tc-99m disphosphonate and Tc-99m pyrophosphate,[26,27] and it was felt that hypervascularity and microscopic calcification in the tumor contributed to the uptake of the radioactivity. Enneking et al.[28] have also demonstrated that on the radionuclide bone scan, increased uptake in the bone adjacent to a soft tissue sarcoma indicates bone involvement that may be critical to planning the appropriate resection. Mori et al.[29] found that Tc-99m bleomycin scans detected 80 percent of various malignant tumors, including seven of eight "cancer and sarcoma" of the extremities. Kim et al.[30] also showed positive imaging of three liposarcomas with inhalation of Xe-133 gas which is highly soluble in fat.

REFERENCES

1. Gilday DL, Ash JM: Benign bone tumors. Sem Nucl Med 1976; 6:33–46
2. Martin NL, Preston DF, Robinson RG: Osteoblastoma of the axial skeleton shown by skeletal scanning. J Nucl Med 1976; 17:187–189
3. Smith FW, Gilday DL: Scintigraphic appearance of osteoid osteoma. Radiol 1980; 137:191–195
4. Humphry A, Gilday DL, Brown RG: Bone scintigraphy in chondroblastoma. Radiol 1980; 137:497–499
5. Brenner RJ, Hattner RS, Lilien DL: Scintigraphy features of nonosteogenic fibroma. Radiol 1979; 131:727–730
6. Siddiqui AR, Tashjian JH, Lazarus K, et al: Nuclear medicine studies in evaluation of skeletal lesions in children with histiocytosis X. Radiol 1981; 140:787–789
7. McNeil BJ, Hanley J: Analysis of serial radionuclide bone images in osteosarcoma and breast cancer. Radiol 1980; 135:171–176
8. Goldstein H, McNeil BJ, Zufall E, et al: Changing indications for bone scintigraphy in patients with osteosarcoma. Radiol 1980; 135:177–180
9. Goldman AB, Braunstein P: Augmented radioactivity on bone scan of limbs bearing osteosarcomas. J Nucl Med 1975; 16:423–426
10. Enneking WF, Kagan A: "Skip" metastases in osteosarcoma. Cancer 1975; 36:2192–2195
11. Brower AC, Teates CD: Positive ^{99m}Tc-polyphosphate scan in case of metastatic osteogenic sarcoma and hypertrophic pulmonary osteoarthropathy. J Nucl Med 1974; 15:53–57
12. McNeil BJ, Cassady JR, Geiser CF, et al: Fluorine-18 scintigraphy in children with osteosarcoma or Ewing's sarcoma. Radiology 1973; 109:627–632
13. Bigler RE, Rosen G, Tofe AJ, et al: Distribution of ^{32}P-disphosphonate in patients with osteogenic sarcoma, in Proceedings of American Association for Cancer Research, Toronto, Canada, 1976
14. Jenkin RDT: Ewing's sarcoma: A study of treatment methods. Clin Radiol 1966; 17:97–106
15. Frankel RS, Jones AE, Cohen JA, et al: Clinical correlations of ^{67}Ga and skeletal whole body radionuclide studies with radiography with Ewing's sarcoma. Radiol 1974; 110:597–603
16. Goldstein H, McNeil BJ, Zufall E, et al: Is there still a place for bone scanning in Ewing's sarcoma? J Nucl Med 1980; 21:10–12
17. Woolfenden JM, Pitt MJ, Durie BGM, et al: Comparison of bone scintigraphy and radiography in multiple myeloma. Radiol 1980; 134:723–728
18. Siddiqui AR, Oseas RS, Wellman HN, et al: Evaluation of bone marrow scanning with Tc-99m sulfur colloid in pediatric oncology. J Nucl Med 1979; 20:386–397
19. Gilbert EH, Earle JD, Glatstein E, et al: ^{111}Indium bone marrow scintigraphy as

an aid in selecting marrow biopsy sites for the evaluation of marrow elements in patients with lymphoma. Cancer 1976; 38:1560–1567

20. Dibos PE, Judisch JM, Spaulding MB, et al: Scanning the reticuloendothelial system in hematological diseases. Johns Hopkins Med J 1972; 130:68–73

21. Kniseley RM: Marrow studies with radiocolloids. Sem Nucl Med 1972; 2:71–81

22. Thompson DE, Frost HM, Hendrick JW, et al: Soft tissue sarcomas involving the extremities and the limb girdles: a review. South Med J 1971; 64:33–44

23. Kaufman JH, Cedermark BJ, Parthasarathy KL, et al: The value of [67]GA scintigraphy in soft tissue sarcoma and chondrosarcoma. Radiol 1977; 123:131–134

24. Bitran JD, Beckerman C, Golomb HM, et al: Scintigraphic evaluation of sarcomata in children and adults by Ga-67 citrate. Cancer 1978; 42:1760–1765

25. Pinsky SM, Henkin RE: Gallium-67 tumor scanning. Sem Nucl Med 1976; 6:397–408

26. Pearlman AW: Preoperative evaluation of liposarcoma by nuclear imaging. Clin Nucl Med 1977; 2:47–51

27. Blatt CJ, Hayt DB, Desai M, et al: Soft tissue sarcoma: Imaged with Tc-99m pyrophosphate. NY State J Med 1977; 77:2118–2119

28. Enneking WF, Chew FS, Springfield DS, et al: The role of radionuclide bone scanning in determining the resectability of soft tissue sarcomas. J Bone Joint Surg 1981; 63:249–257

29. Mori T, Hamamoto K, Onoyama Y, et al: Tumor imaging after administration of ^{99m}Tc-labeled bleomycin. J Nucl Med 1975; 16:414–422

30. Kim EE, Deland FH, Maruyama Y, et al: Detection of lipoid tumors by Xe-133. J Nucl Med 1978; 19:64–66

CHAPTER 10

Endocrine Tumors

INTRODUCTION

The vitality of nuclear medicine is reflected by the fact that the diagnosis and treatment of thyroid disease with radionuclides is the oldest area of clinical endeavor and yet continues to be the subject of active research, with the development and clinical application of new radiopharmaceuticals and new imaging techniques in the last decade. Ultrasound has become an important adjunctive procedure that has enhanced the value of radionuclide scans by aiding in the distinction between cystic and solid nodules. In the same interval, the efficacy and safety for I-131 therapy for thyroid carcinoma has been further substantiated.

Nuclear medicine also plays a direct role in furthering the understanding of the various multiple endocrine adenomatosis syndromes. Development of radioassays for calcitonin and provocative tests to enhance its secretion have provided the tools necessary to detect and follow patients at risk for medullary thyroid carcinoma.

Radionuclide parathyroid imaging is rarely performed, since parathyroid adenomas are not encountered commonly in clinical medicine, and the available radiopharmaceuticals are less than optimal. However, once the parathyroid hormone (PTH) level is established as elevated by radioassay, location of an adenoma is difficult, and often involves extensive neck and mediastinal dissection. However, neck vein catheterization and measurement of PTH is a valuable tool for locatization of a parathyroid tumor especially if neck exploration fails. Without a palpable mass in the neck, it still may be clinically reasonable in very unusual circumstances to perform parathyroid imaging, even with a high rate of false negatives.

In many respects, radionuclide adrenal imaging is at the other end of the spectrum from the thyroid, being one of the least available procedures in nuclear medicine. Although radiolabeled cholesterol is still an investigational new drug from a regulatory standpoint in the United States, scintigraphic patterns have been defined for numerous conditions, and the clinical utility of adrenal imaging is now considered well established.

With the increasing emphasis on the functional aspects of nuclear imaging, pituitary, parathyroid, and pancreatic endocrine function imaging remain challenges for nuclear medicine that may be met with development of appropriate radiolabeled enzyme inhibitors or highly specific radiolabeled antibodies.

CHARACTERISTICS OF ENDOCRINE TUMORS

Epidemiology
In 1975,[1] the age-adjusted incidence rates for malignant tumors of the thyroid gland in the United States was 2.2 and 5.2 per 100,000 for white males and females, respectively. The age-specific rates increase with age. Geographic areas of endemic goiter, where carcinoma of the thyroid has been more frequent, are disappearing with the general use of iodized salt. Knowlson[2] found eight carcinomas (4.2 percent) in a series of 191 patients with solitary nodules, but only three (1.1 percent) in those with multinodular goiter. Carcinoma of the thyroid is rarely found in patients with diffuse toxic goiter, or in those with toxic nodular goiter. There is evidence that irradiation to the thyroid in early life results in changes that may lead to the development of thyroid cancer.[3] The mean interval between irradiation and the diagnosis of cancer, in a series of such patients, was 10.9 years.[4] In 468 children with carcinoma of the thyroid, 73 percent had received prior irradiation.[5] More recently, about 25 percent of persons given x-ray therapy to the head, neck, or upper chest as children have developed colloid nodular goiter, and 5 percent developed thyroid cancer.[6]

Tumors of the parathyroid gland are rare. Most tumors of the parathyroid are adenomas. Only about 1.5 to 5 percent of these tumors are considered as carcinomas. Patients with parathyroid tumors range in age from 13 to 92 years. In recent years a greater proportion of the tumors have been found in relatively young patients.[7]

Tumors of the adrenal gland occur infrequently. The incidence rate of carcinoma of the adrenal cortex is about 0.2 per 100,000.[8] Most of the functional tumors, benign or malignant, are found in female patients. Cortical adenomas are commonly found at autopsy or in patients with Cushing's syndrome. Benign tumors of the medulla are rare. Neuroblastomas are the most common malignant tumors of the adrenal gland and constitute one-tenth of all tumors seen in childhood.[9] Many are present at birth. Pheochromocytomas have a peak incidence in the fifth decade.[10]

Pathology and Metastasis
The following histologic classification of thyroid cancer has been used:

Adenoma—Follicular, oxyphil (Hürtle cell), atypical.
Carcinoma (differentiated)—Papillary, follicular, mixed papillary-follicular, oxyphil.
Carcinoma (undifferentiated)—Small cell, spindle cell, and giant cell.
Medullary (solid) carcinoma.
Others—Squamous cell carcinoma, malignant lymphoma, plasmacytoma, and sarcoma.

The adenoma is the most common of the thyroid tumors. The majority of carcinomas are mixed, papillary and follicular, with one or the other predominating. Oxyphil carcinomas (Hurtle cell) make up 5 to 10 percent of thyroid carcinomas. Small cell carcinomas are uncommon, and are usually malignant lymphomas. Spindle and giant cell carcinomas are anaplastic, and these tumors may be related to preexisting papillary or follicular carcinoma.

Medullary carcinomas are unusual, and there is evidence to support their pathogenesis from the parafollicular or C cell.[11] Squamous cell carcinomas constitute less than 1 percent of thyroid carcinomas. Primary malignant lymphomas, plasmacytomas, and sarcomas are rare. Multicentric foci of thyroid carcinoma within the gland are frequently reported, and these lesions are usually considered as multiple primary carcinomas rather than metastases. Regional lymph node metastases are found, in the majority of cases, in the jugular, supraclavicular and mediastinal chains. The pretracheal and paratracheal lymph nodes are the first involved. The reported proportion of tumors presenting lymph node metastases varies with the histologic type; 2 to 13 percent in a follicular carcinoma, and 40 to 70 percent in papillary carcinomas.[12] Medullary carcinomas have a better than 50 percent chance of metastasis, particularly those of the intermediate and fibrotic types. Organs affected by metastases are lungs, bone, liver, kidneys and brain. Pulmonary metastases are usually subpleural and multiple. The pelvic bones, the vertebrae, and the sternum were the most common sites of bony metastases found contemporaneously with the primary tumor of the thyroid, and 47 (52 percent) were observed from 6 months to 20 years after treatment of the primary lesion.[13]

Half of the parathyroid tumors seem to rise from the lower parathyroid glands, on the dorsal surface of the lower lobes of the thyroid.[14] Nathaniels, et al.[15] reported a series of 84 parathyroid tumors of the mediastinum, 72 of which were in the anterior portion of the upper mediastinum.

Adenomas may contain water-clear cells that should not be confused with water-clear cell hyperplasia that involves all four parathyroid glands. Carcinomas are distinguished from adenomas by the presence of mitotic figures, vascular invasion, and evidence of metastases.

Nearly one-third of all cases of the parathyroid carcinoma are found to have metastasized to regional lymph nodes, lungs, liver, and bones.[16]

Tumors that develop from the adrenal cortex have an epithelial character, whereas those that develop from the medulla are nervous sytem tumors. Tumors can be classified as follows:

Tumors from adrenal cortex—Adenoma (nonfunctioning, functioning) carcinoma.

Tumors from adrenal medulla—Ganglioneuroma, pheochromocytoma, neuroblastoma functioning.
Others—Lipomas, myomas, angiomas, fibromas, fibrosarcomas, metastatic
tumors.

Most of the cortical tumors are benign. The cortical adenoma is frequently found at postmortem examination, and it is usually 2 cm or less in diameter.
Functioning tumors of the aldosterone-secreting type are usually benign. One
in 14 of these tumors may be malignant.[17] Carcinomas of the adrenal cortex
occur more often on the left than on the right side and are usually larger than
benign tumors, measuring 4 to 15 cm in diameter. Tumors arising from the
medulla are of nerve origin, and ganglioneuromas are usually found by
chance. Pheochromocytomas are frequently bilateral and may be multiple. A
positive Henle chromoreaction is a valuable help for the diagnosis. Remine et
al.[18] found 18 malignant tumors among 138 cases. About 20 percent of these
tumors may be found in ectopic situations, including the chest and the
bladder.[19] Neuroblastomas are usually small, but may reach a large size. They
invariably show areas of hemorrhage and necrosis.

Carcinomas of the adrenal cortex metastasize predominantly to the liver,
lungs, brain, and regional lymph nodes, but rarely to bones. Malignant
pheochromocytomas metastasize to those same organs and areas, but metastases to bones are more common.[18] Some neuroblastomas characteristically
metastasize massively to regional lymph nodes and liver. Other neuroblastomas may present clinically with metastases to the skull; bone metastases may
also occur in the sternum, vertebrae, ribs, and long bones, apparently utilizing
the vertebral vein plexus.

Clinical Evolution

Most thyroid tumors develop slowly, sometimes over several decades. The
first sign is usually an enlargement or nodule. Hoarseness due to laryngeal
displacement or paralysis may be found. Papillary carcinomas, which occur
frequently in young patients, have the most benign course; lymph node
metastases may be an initial sign. Follicular carcinomas are usually large
when first seen, and the patient may already present bony metastases. Lung
metastases are next in frequency, and may be found in as many as 6 percent of
all presenting cases.[20]

A large proportion of patients with parathyroid adenomas do not present
clinical symptoms or an ostensible mass. Many of them are incidentally
discovered, and about 25 percent are discovered only at autopsy.[14] But in
about 50 percent of the patients, the first clinical symptoms are those of
hyperparathyroidism: malaise, fatigue, and weakness. Others may present
bone lesions: bone cysts and osteoporosis. Carcinomas of the parathyroid also
may be discovered during investigation of nephrolithiasis, bone lesions, or
hypercalcemia, but in 30 to 70 percent the primary tumor becomes palpable,
an important differential point with cases of adenoma and hyperplasia.[7]

The majority of adrenal cortical adenomas do not produce symptoms. A
small proportion of these tumors may cause virilizing alteration in females
after puberty and before menopause. Less frequently, a feminizing tendency
may be observed in males.

Tumors, as well as hyperplasia of the adrenal cortex, may be associated with Cushing's syndrome. In a series of 110 patients with Cushing's syndrome, 26 percent had adrenocortical tumors, one-third of which were carcinomas.[21] Patients with aldosterone-producing tumors may have numerous complaints, including muscle weakness and polyuria, and suffer the consequences of hypokalemia and hypertension. Cortical carcinomas may cause hormonal alterations as variable as those of adenomas. Most patients have no impressive symptoms: asthenia, low-grade fever, loin pain. Ganglioneuromas rarely produce any symptoms other than those caused by their increasing size. Pheochromocytomas may present a very dramatic clinical evolution. The tumor may cause paroxysmal attacks due to intermittent flooding of the bloodstream with pressor substances. Headache, perspiration, palpitation, pallor, nausea, tremor, weakness, and anxiety may be recognized. Neuroblastomas cause varied and often vague symptoms. The Pepper type is characterized by anemia, weakness, weight loss, and abdominal distention with hepatomegaly. The Hutchinson type is characterized by ecchymosis of the eye with proptosis, and enlargement of the preauricular, submaxillary, and upper cervical lymph nodes.

NUCLEAR IMAGING

Imaging of Primary Tumors

Thyroid Imaging. The most frequent indication for thyroid scanning is the evaluation of clinically palpable abnormalities. It is important to determine whether the nodule is functioning and whether the nodule is solitary or multiple (Fig. 1). These considerations largely determine patient management.

Thyroid nodules are common. In as many as 50 percent of thyroid autopsies, the gland has nodules that have not been suspected clinically, and in 75 percent of these cases multiple nodules are found. The incidence of nodules increases with age, and both benign and malignant tumors are more common in women than in men. Atkins[22] reviewed several studies reporting the incidence of malignancy in thyroid nodules. In over 2,000 solitary nodules thyroid cancer was found in 2.1 percent of hot nodules, 40 percent of warm nodules, and 19.8 percent of cold nodules. Nonfunctioning cold nodules represent adenoma, carcinoma, cyst, hemorrhage, edema, fibrosis, and thyroiditis. Thirty-seven children and adolescents with a solitary thyroid nodule were seen over a 16-year period and 35 had their nodules removed surgically.[23] All patients had preoperative scans, of which 27 showed a cold nodule (Fig. 2). The most common cause of solitary thyroid nodule was follicular adenoma. Five of 27 cold nodules were malignant (18.5 percent) while no malignancies were present in the warm or hot nodules. Cold nodules that involve an entire lobe are more apt to be due to thyroiditis. Large cold nodules with smooth borders are more often benign cysts (Fig. 3). Nodules associated with hyperthyroidism or calcification have a low incidence of malignancy.

In general, surgery is recommended in a solitary nodule in women under

Figure 1. Thyroid image with Tc-99m pertechnetate shows hot nodules in the right and left lobes.

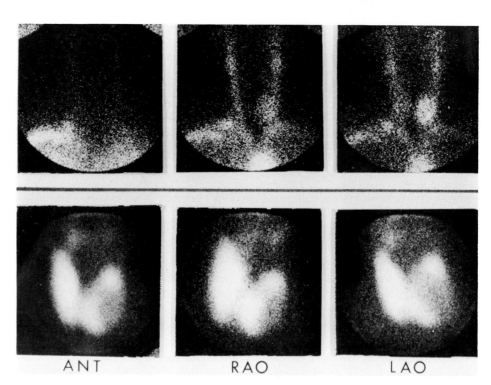

ANT   RAO   LAO

Figure 2. Routine views of the thyroid (lower row) with Tc-99m pertechnetate show a large cold nodule in the left lobe. Dynamic flow study (upper row) suggests a vascular lesion. A follicular adenoma on biopsy.

Figure 3. Tc-99m thyroid image shows a single cold nodule in the upper portion of the left lobe. A cyst on ultrasonography.

40, solitary nodules in males at any age, a nodule that fails to decrease after several months of thyroid suppression, nodules associated with palpable lymph nodes, and multinodular glands in males under 40 who do not live in a goitrous area and who have not had a history of thyroiditis. The suspicion of thyroid carcinoma in patients with multinodular goiter is very much lower than in patients with solitary nodules. DeGroot and Stanbury[24] point out that only 25 new thyroid tumors appear per year per million population.

The risk of thyroid cancer to individuals irradiated during childhood is great. Arnold et al.[25] examined 1,452 persons who had documented or presumed irradiation to the neck region for benign conditions 18 to 35 years previously. Thyroid abnormalities were found in 21 percent of these patients. Of those patients who underwent surgical exploration, 29 percent were found to have thyroid cancer. The overall incidence of thyroid cancer in these neck-irradiated patients was 7 percent.

Although Tc-99m pertechnetate and I-123 used in combination with the scintillation camera and pinhole collimator provide images of superb technical quality, on the basis of cost and availability, Tc-99m pertechnetate still appears to be the agent of choice for clinical use. However, there has been a number of reports describing discordant scan results (Fig. 4) when Tc-99m pertechnetate images have been compared with those obtained with I-131 or I-123 in the same patients.[26] Some of these abnormalities ultimately proved to be thyroid carcinomas.[27] The putative mechanism for the discordant findings is an uncoupling of the trapping mechanism from the organification mechanism in these nodules.

While the papillary elements usually do not concentrate iodine, there are very few pure papillary thyroid cancers. Most contain some follicular ele-

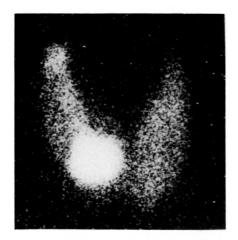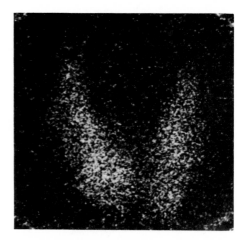

Figure 4. Discordant thyroid images with Tc-99m and I-131. A hot nodule seen only on Tc-99m image was a Hürtle cell adenoma on biopsy.

ments and therefore concentrate radioiodine to some extent (Fig. 5). Differentiated tumors concentrate iodine better than undifferentiated and Hürtle cell cancers.

Pretreatment evaluation of patients with thyroid carcinoma can be done utilizing standard techniques. The use of Tc-99m pertechnetate or I-123 in combination with the pinhole collimator will allow the detection of some thyroid abnormalities that are too small to be clinically palpable. This is particularly important in assessing patients who are at a high risk for thyroid carcinoma.

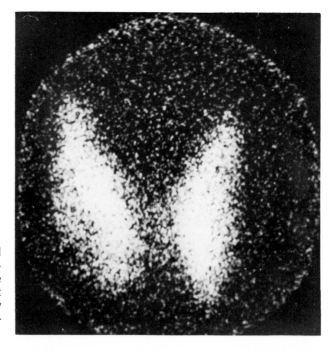

Figure 5. I-131 thyroid image shows an ill-defined cold nodule in the lateral aspect of the right inferior pole. Biopsy shows a papillary-follicular mixed carcinoma.

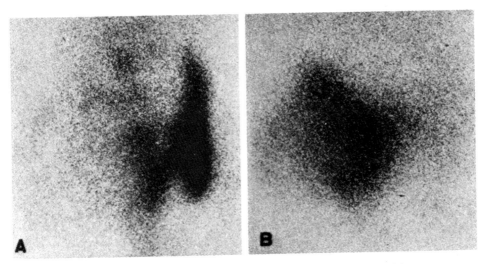

Figure 6. A hard mass palpated in the right neck of a 72-year-old man. (a) Tc-99m thyroid image shows a large cold nodule in the right lobe. (b) Tl-201 image reveals a remarkable uptake of activity corresponding to the cold nodule. Surgery found a papillary adenocarcinoma. *(From Radiol 1978; 129:499, with permission.)*

The quantification of total thyroidal iodine by the fluorescent scan using Am-241 has been applied to the differential diagnosis of thyroid nodules. The rationale for this approach is that thyroid malignancies contain low concentrations of iodine, and many benign lesions causing scintigraphically cold nodules have significant iodine concentrations.[28]

Employing early and delayed T1-201 scans (Fig. 6), it has been possible to differentiate malignant thyroid tumors from those that were benign. Seventy-six cases of histologically verified thyroid tumors, all seen as cold nodules on the I-123 thyroid scans, were imaged 5 to 10 minutes (early) and 3 to 5 hours (delayed) after intravenous injection of 1-2 mCi T1-201. In 35 (94.6 percent) of 37 malignant tumors, T1-201 accumulated in the cold nodule of the I-123 thyroid scan on early and delayed images. On the other hand, the delayed T1-201 scan was negative in 35 out of 39 (89.7 percent) benign tumors.[29] Siddiqui et al.[30] also used Tc-99m diphosphonate imaging in the differential diagnosis of thyroid nodules. All cystic lesions had less uptake of Tc-99m diphosphonate as compared with the uptake in normal. Forty-eight of fifty solid nodules had the uptake to the same or greater degree than did the normal thyroid gland, but both carcinoma and adenoma had identical appearance on the scans.

The scintigraphic differentiation of the malignant from the benign thyroid cold nodule had also been attempted using the Ga-67 citrate.[31] The uptake of Ga-67 usually has been in poorly differentiated or anaplastic carcinoma (Fig. 7), and no benign nodule has been reported to concentrate Ga-67 although uptake has been observed in patients with thyroiditis.[32]

Parathyroid Imaging. The greatest difficulty encountered in clinical parathyroid scanning is the frequency of false negative examinations. Thus far, there has been no reliable way to augment parathyroid uptake of Se-75 selenometh-

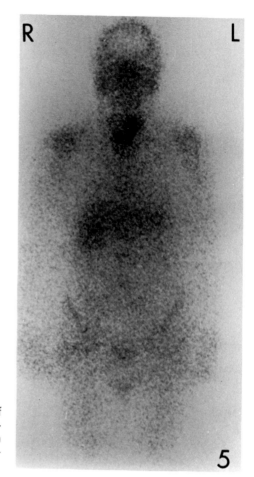

Figure 7. Longitudinal section image of whole body 48 hours following the injection of Ga-67 citrate shows radiogallium uptake in the recurrent poorly differentiated thyroid carcinoma.

ionine in the clinical setting. The site of the abnormal parathyroid gland was identified in 50 percent of the total 50 patients or 60 percent of those with active disease.[32] Patients with more severe biochemical disease tended to have more positive scans, suggesting that the ability to demonstrate the gland is related to overall glandular activity.

Benign adenomas are rarely multiple, and most adenomas develop from the inferior parathyroid gland. Approximately 10 percent of adenomas are not readily identified at surgery, but this varies with the experience of the surgeon. Multiple adenomas are usually associated with a multiglandular endocrinopathy.

There has been a recent report of preoperative localization of a mediastinal parathyroid adenoma with Tc-99m pertechnetate scan.[33] Pertechnetate may have been overlooked as a useful imaging agent for parathyroid adenomas, especially for tumors located away from the thyroid (Fig. 8).

Adrenal Imaging. Adrenal scintigraphy has been clinically feasible since the development of I-131-19-iodocholesterol in 1970. This agent has been supplanted by the current I-131-6-iodomethyl-19-norcholesterol. Patients receive

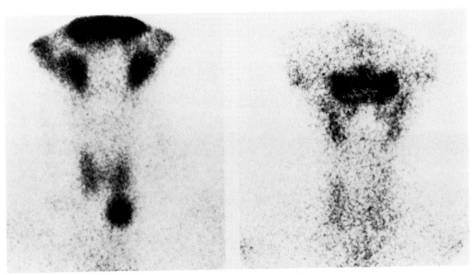

Figure 8. Preoperative Tc-99m thyroid scan (left) shows abnormal uptake in mediastinum at site of parathyroid adenoma. Follow-up scan (right) reveals normal thyroid image without ectopic uptake following removal of parathyroid adenoma. *(From J Nucl Med 1982; 23:512, with permission.)*

Lugol's solution to block the thyroid gland and receive 1-2 mCi of radiocholesterol intravenously. Imaging is accomplished 4 to 7 days postinjection with the gamma camera. In normal adrenal scintigraphy, the right adrenal gland is higher than the left and appears slightly hotter. The left adrenal has an oval configuration, while the right adrenal has a truncated or circular configuration in most subjects.

Adrenal percent uptake determination is similar to thyroid uptakes and may be accomplished with the aid of a digital computer and standard percent uptake curves derived from phantom studies. In a series of normal subjects studied with 6-iodomethyl-19-norcholesterol, the mean percent per gland uptake was 0.164 percent, with a range of 0.073 to 0.259 percent.[34] When individual depth corrections were made, the greatest right to left discrepancy was 16 percent and right to left ratios ranged from 0.94 to 1.16.

With documented glucocorticoid excess, symmetrical visualization is due to adrenal hyperplasia, usually secondary to Cushing's disease. Unilateral visualization indicates the presence of an adenoma (Fig. 9) or a postsurgical adrenal remnant; and bilateral nonvisualization is typically due to carcinoma.

Adrenal suppression scans were developed to enhance differences between the normal and abnormal adrenal cortex in certain clinical conditions. Patients receive dexamethasone prior to radionuclide injection, and serial scans beginning 2 to 3 days postinjection are obtained. On suppression scans in primary aldosteronism and adrenal androgenism, adenomas demonstrate unilateral or markedly asymmetrical uptake (Fig. 10). Patients with micro and macronodular hyperplasia typically demonstrate bilateral uptake, whereas in normal subjects there should be no visualization while on dexamethasone suppression.

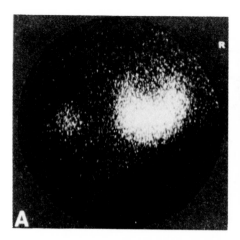
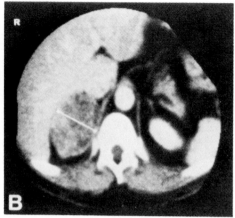

Figure 9. After 11 days of ACTH, preoperative posterior adrenal scan (**a**) at 6 days after I-313 19-iodocholesterol injection shows marked uptake of activity in right corticoadrenal adenoma and scintigraphic reactivation of suppressed left adrenal gland. CT (**b**) shows right adrenal mass. *(From J Nucl Med 1981; 22:1059, with permission.)*

Distinctly lateralizing suppression scans have a 94 percent specificity for aldosteronoma and are felt to represent sufficient evidence for this diagnosis—that adrenal venous catheterization is not necessary prior to surgery when this pattern is demonstrated.[35]

Although radioiodinated cholesterols do not localize in medullary tissue, adrenal medullary disorders such as pheochromocytoma may be diagnosed by

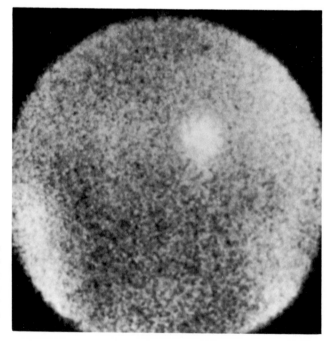

Figure 10. Posterior adrenal image at 5 days after I-131 6-NP injection with continuous suppression shows an aldosterone-producing right adrenal adenoma.

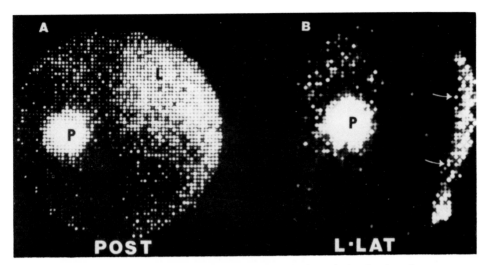

Figure 11. Posterior and left lateral views of adrenal gland 1 day after I-131 MIBG injection shows remarkable uptake of activity in left pheochromocytoma (P). L: liver; arrows: patient's back marker. *(From N Engl J Med 1981; 305:15, with permission.)*

displaced, distorted, or destroyed cortical tissue. Such lesions must be 2 cm in diameter or larger to be detected. Recently, scintigraphy with I-131 meta-iodobenzyl guanidine (Fig. 11) has localized hyperfunctioning adrenergic tissue in patients with pheochromocytoma.[36] The results document that scintigraphy using I-131 meta-iodobenzyl guanidine seems to be a safe and reliable method of detecting adrenal and extra-adrenal pheochromocytomas, both benign and malignant. Radionuclide bone scan may show a radioactivity uptake in necrotic neuroblastomas (Fig. 12).

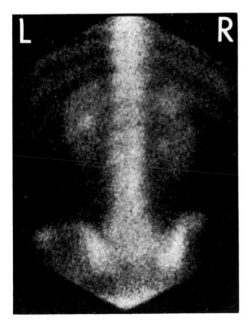

Figure 12. Posterior abdominal image with Tc-99m MDP shows radioactivity accumulating at the site of necrotic right neuroblastoma.

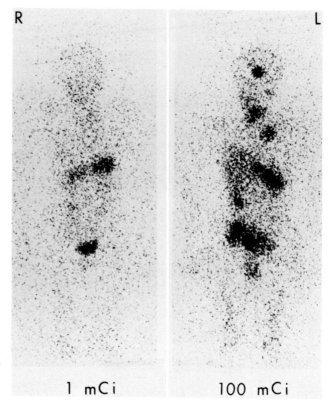

Figure 13. Anterior whole-body images with 1 and 100 mCi I-131 show recurrent follicular thyroid carcinoma with metastasis in the left upper lung detected only on high-dose scan.

Imaging of Metastatic Tumors

For the postoperative diagnosis and management of thyroid carcinoma metastasis, the first step is ablation of any remaining thyroid gland tissue. The preferred means of ablation is surgical, since it removes the greatest amount of tissue from the neck and induces rapid onset of hypothyroidism and elevated TSH. After the initial thyroidectomy, patients are not started on thyroid replacement for a period of 6 weeks. At the end of this 6-week interval, extended field scans (Fig. 13) are obtained using 1-10 mCi. If areas of residual thyroid carcinoma are seen in this study, an ablating dose (100 to 150 mCi) of radioiodine is administered before the patient is begun on thyroid replacement.

Subsequent scans are obtained at 1, 3, and 5 years, and every 5 years after that, assuming that no new disease is detected. On these later scans, the patient discontinues all thyroid replacement medication for a period of 6 weeks, which will usually produce symptoms of hypothyroidism. This method appears to be much more satisfactory than the administration of TSH, and obviates the risk of immunologic reactions to repeated injections of foreign protein. Another method that may be used to detect metastasis of thyroid cancer is to scan using Tc-99m pertechnetate, which may be useful in tumors that trap but do not organify iodine. It is possible that some metastases may be detected in this way, although this technique should always be used as a secondary measure because of high background activity and no intimation of the metastases' response to therapeutic iodine.

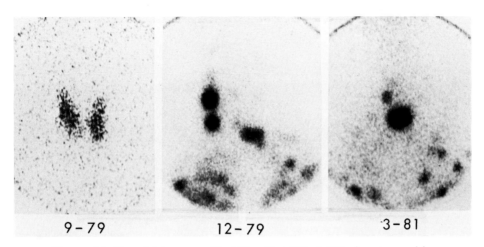

9-79   12-79   3-81

Figure 14. Thyroid image with 100 uCi I-131 (9-79) shows a cold nodule in the right inferior pole. Biopsy revealed a papillary thyroid carcinoma. Follow-up images of neck and 2 mCi (12-79, 3-81) after total thyroidectomy show metastases in cervical and paraclavicular lymph nodes.

Most metastases of thyroid cancer are found in the lymph nodes; the next most common sites are the lung and bone (Figs. 14 to 16). Tc-99m phosphate bone scans were positive for metastasis in nine of 32 patients with medullary thyroid carcinoma. These patients had elevated serum thyrocalcitonin levels.[37] Tc-99m phosphate bone scans can localize metastatic bone foci that do not concentrate radioiodine.[38]

There has been a report of detection of metastatic adrenal carcinoma using I-131-6-β-iodomethyl-19-norcholesterol total-body scan.[39] Scintigraphy using I-131 meta-iodobenzyl guanidine also locates extra-adrenal pheochromocytoma and thereby directs the surgeon's exploration.[40]

Treatment of Thyroid Cancer with Radioactive Iodine

I-131 treatment of well differentiated thyroid carcinoma is a well-evaluated therapeutic model for nuclear medicine that has never been equaled by subsequent developments. It is still a unique method of treating cancer, since I-131 irradiates the metastatic lesion from the inside out relatively selectively (Fig. 17).

Data has been accumulated to show that well differentiated thyroid cancer does kill commonly enough to warrant aggressive treatment with surgery and radioiodine, even in young patients.[41] There is also data demonstrating that I-131 after surgery decreases the recurrence rate and death rate from well-differentiated thyroid cancer. The modality of I-131 and surgical treatment is reasonable in contrast to the recurrence and death rate from nonaggressively treated well differentiated thyroid carcinoma. There has been no decrease in fertility and no increased incidence of birth abnormality or leukemia following I-131 therapy for thyroid carcinoma even in children.[41]

I-131 does not produce significant pulmonary fibrosis with deterioration of lung function, unless the patient has an uptake of 100 mCi or more in a single dose in lungs.

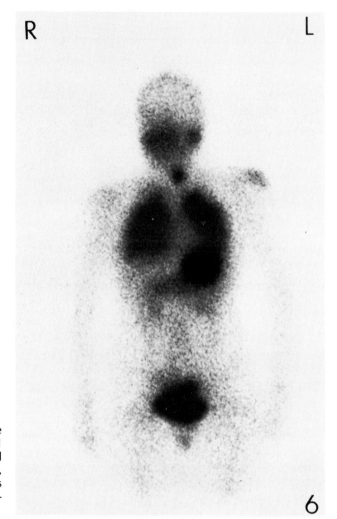

Figure 15. Longitudinal section image of whole body with 2 mCi I-131 shows diffuse abnormal activities in both lungs, consistent with metastasis of recurrent follicular thyroid carcinoma.

OTHER RADIOLOGIC IMAGING

Ultrasonography

The differentiation of cyst from adenoma or carcinoma of the thyroid cannot be made on the basis of a static radionuclide thyroid scan alone. Gray scale ultrasound with specialized high frequency (5MHz), small diameter, internally focused transducers can reliably distinguish between cystic and solid lesions of the thyroid. Patients with scintigraphically cold nodules, which are cystic by ultrasound, can be treated conservatively or with needle aspiration. Those lesions that demonstrate a solid or mixed solid cystic structure by ultrasound evaluation can be referred for more definitive therapy.

Radionuclide subtraction (Se-75-Tc-99m) technique and high resolution gray-scale echography provided complementary information in   reoperative assessment of parathyroid lesions and facilitated localization   .eno-mas well beyond the resolution of other reported techniques.[42]

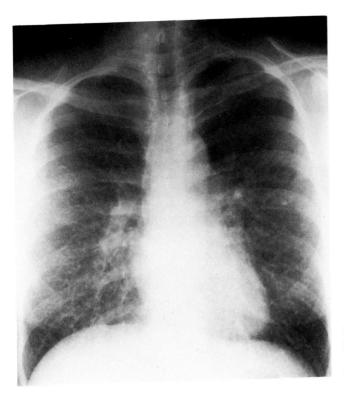

Figure 16. Chest radio-graph (same case as Fig. 15) appears essentially clear.

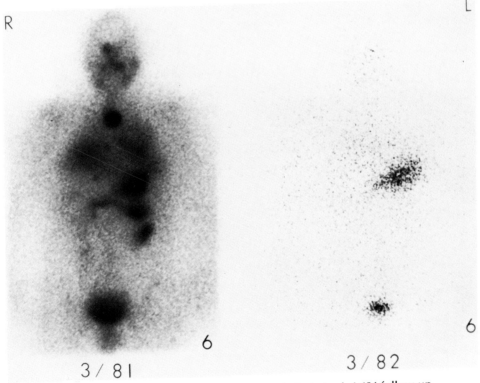

R   L

3 / 81   3 / 82

Figure 17. Longitudinal section images of whole-body I-131 follow-up scans show recurrent follicular thyroid cancer with lung metastasis (3/81) and their complete resolution 1 year after 200 mCi I-131 treatment (3/82).

The ultrasonic appearance of adrenal tumors, except for cyst or myeloli-poma, is relatively nonspecific and can be produced by adrenal metastases.[43] Besides the technical expertise required, a major difficulty with ultrasound is the high percentage of nondiagnostic cases secondary to overlying bowel gas or obesity of the patient.

Computed Tomography

CT is now the most sensitive means of noninvasively detecting adrenal masses or diffuse enlargement.[44] Except for adrenal cysts and myelolipomas, primary adrenal tumors are difficult to distinguish from secondary tumors by CT. No significant information about thyroid imaging by CT is available, but better delineation of thyroid nodules that lie at different depths within the thyroid gland can be expected from tomographic reconstruction. Emission computed axial tomography (ECAT) should permit accurate quantitation of the total volume of the thyroid gland and abnormalities within it, and it should also permit accurate regional quantitation of radioactivity within normal and abnormal portions of the thyroid gland.

Angiography

Adrenal venography was available clinically for a decade prior to adrenal scintigraphy. However, venography with aldosterone determinations (unsuc-cessful on the right in 10 to 36 percent), is no more sensitive than adrenal scintiscan, and is actually less specific in differentiating macronodular hyper-plasia from adenoma.[35] This is not surprising in that many macronodules are actually larger than some adenomas and can result in a completely identical radiographic appearance.

Arteriography and lymphography for the diagnosis of thyroid cancer have not been widely developed.

Radiography

X-ray examination of the neck disclosing tracheal deviation or the fine stippled calcification of psammoma bodies may suggest thyroid malignancy, whereas broad bands or rings of calcification are usually found in benign thyroid lesions.

REFERENCES

1. Culter SJ, Young JL Jr.: Third national cancer survey: Incidence data. Natl Cancer Inst Monogr 1975; 41:10–27, 100–135, 388–427
2. Knowlson GTG: The solitary thyroid nodule. Br J Surg 1971; 58:253–254
3. Hempelmann LH: Risk of thyroid neoplasms after irradiation in childhood. Science 1968; 160:159–163
4. Raventos A, Winship T: The latent interval for thyroid cancer following irradia-tion. Radiol 1964; 83:501–508
5. Winship T, Rosvoll RV: Cancer of the thyroid in children. Proc Natl Cancer Conf 1970; 6:677–681
6. Refetoff S, Harrison J, Jaranfilski BT, et al: Continuing occurrence of thyroid carcinoma after irradiation to the neck in infancy and childhood. N Eng J Med 1975; 292:171–173

7. Schantz A, Castleman B: Parathyroid carcinoma. Cancer 1973; 31:600–605

8. Lubitz JA, Freeman L. Okum R: Mitotane use in inoperable adrenal cortical carcinoma. JAMA 1973; 223:1109–1112

9. Louw JH: Flank tumors in infants and children. S Afr J Surg 1974; 12:5–18

10. Remine WH, Chong GC, Van Heerden JA, et al: Current management of pheochromocytoma. Ann Surg 1974; 179:740–748

11. Ljungberg O: Argentaffin cells in human thyroid and parathyroid and their relationship to C-cells and medullary carcinoma. Acta Pathol Microbiol Scand 1972; 80:589–599

12. Crile G, Jr., Hawk WA: Carcinomas of the thyroid. Cleve Clin Q 1971; 38:97–104

13. Dargent M, Colon J, Lahneche B, et al: Les metastases osseuses du cancer thyroidien. Ann Radiol 1970; 13:483–506

14. Krementz ET, Yeager R, Hawley W, et al: The first 100 cases of parathyroid tumor from Charity Hospital of Louisiana. Ann Surg 1971; 173:872–873

15. Nathaniels EK, Nathaniels AM, Wang C: Mediastinal parathyroid tumors: A clinical and pathological study of 84 cases. Ann Surg 1970; 171:165–170

16. Schantz A, Castleman B: Parathyroid carcinoma. Cancer 1973; 31:600–605

17. Harrison JH, Mahoney EM, Bennett AH: Tumors of the adrenal cortex. Cancer 1973; 32:1227–1235

18. Remine WH, Chong GC, Van Heerden JA, et al: Current management of pheochromocytoma. Ann Surg 1974; 179:740–748

19. Fries JG, Chamberlin JA: Extra-adrenal pheochromocytoma: Literature review and report of a cervical pheochromocytoma. Surgery 1968; 63:268–279

20. Staunton MD, Greening WP: Clinical diagnosis of thyroid cancer. Br Med J 1973; 4:532–535

21. Scott HW, Jr.: Tumors of the adrenal cortex and Cushing's syndrome. Proc Natl Cancer Conf 1973; 7:513–527

22. Atkins HL: The thyroid, in Freeman LM, Johnson PM (eds): Clinical Scintillation Imaging, ed 2. New York, Grune and Stratton, 1975

23. Hung W, August GP, Randolph JG, et al: Solitary thyroid nodules in children and adolescents. J Pediat Surg 1982; 17:225–229

24. DeGroot LJ, Stanbury JB: The Thyroid and Its Diseases, ed 4. New York, Wiley 1975

25. Arnold JE, Pinsky SM, Ryo UY, et al: [99m]Tc pertechnetate thyroid scintigraphy in patients predisposed to thyroid neoplasms by prior radiotherapy to the head and neck. Radiology 1975; 115:653–658

26. Kim E, Kumar B: Discrepant imaging of nodular hyperplasia with pertechnetate and radioiodine. Clin Nucl Med 1976; 1:204–205

27. Turner JW, Spencer RP: Thyroid carcinoma presenting as a pertechnetate "hot" nodule, but without ^{131}I uptake. J Nucl Med 1976; 17:22–23

28. Patton JA, Hollifield JW, Brill AB et al: Differentiation between malignant and benign thyroid nodules by fluorescent scanning. J Nucl Med 1976; 17:17–22

29. Ochi H, Sawa H, Fukuda T, et al: Thallium-201 cholorid thyroid scintigraphy to evaluate benign and/or malignant nodules. Usefulness of delayed scan. Cancer 1982; 50:236–240

30. Siddiqui AR, Wellman HN, Park HM, et al: Tc-99m disphosphonate imaging in the differential diagnosis of thyroid nodule. Clin Nucl Med 1982; 7:353–356

31. Kim EE, Maruyama Y, DeLand FH: Substernal thyroid carcinoma detected by Ga-67 scan in a patient with normal I-131 scan. Clin Nucl Med 1978; 3:222–224

32. Potchen EJ, Watts HG, Awward HK: Parathyroid scintiscanning. Radiol Clin N Amer 1967; 5:267–271

33. Naunheim KS, Kaplan EL, Kirchner PT: Preoperative Tc-99m imaging of a substernal parathyroid adenoma. J Nucl Med 1982; 23:511–513

34. Freitas JE, Thrall JH, Swanson DP, et al: Normal adrenal imaging. J Nucl Med 1977; 18:599

35. Seabold JE, Cohen EL, Beierwaltes WH, et al: Adrenal imaging with [131]I-19-iodocholesterol in the diagnostic evaluation of patients with aldosteronism. J Clin Endocrinol Metab 1976; 42:41–46

36. Winterberg B, Fisher M, Better H: Scintigraphy in pheochromocytoma. Klin Wochenschr 1982; 60:631–633

37. Johnson DG, Coleman Re, McCook TA, et al: Radionuclide bone and liver scanning in patients with medullary carcinoma of the thyroid. J Nucl Med 1982; 23:20–24

38. Klinger L: Polyphosphate bone scans, P-32, and adenocarcinoma of the thyroid. J Nucl Med 1974; 15:1037–1038

39. Seabold JE, Haynie TP, Diosdado DN, et al: Detection of metastatic adrenal carcinoma using ^{131}I-6- -iodomethyl-19-norcholesterol total body scans. JJCEM 1977; 45:788–797

40. Sisson JC, Frager MS, Valk TW, et al: Scintigraphic localization of pheochromocytoma. N Eng J Med 1981; 305:12–17

41. Beierwaltes WH: The treatment of thyroid carcinoma with radioactive iodione. Semin Nucl Med 1978; 8:79–94

42. Crocker EF, Jellins J, Freund J: Parathyroid lesions localized by radionuclide subtraction and ultrasound. Radiol 1979; 130:215–217

43. Sample WF: Adrenal ultrasonography. Radiol 1978; 127:461–466

44. Korobkin M, White EA, Kressel HY, et al: Computed tomography in the diagnosis of adrenal disease. Am J Roentgenol 1979; 132:231–238

CHAPTER 11

Hematologic Tumors

INTRODUCTION

Although the prognosis of patients with lymphoma and leukemia has improved markedly in recent years, these diseases represent a major clinical problem. The malignant lymphomas represent a heterogeneous group of disorders of the lymphoreticular system with considerable variation in clinical features, natural history, and outcome after treatment. One of the group, Burkitt's lymphoma, may provide one example of the role of viruses in the etiology of human malignancies, as well as demonstrating the role of immunologic host defenses in cancer. Hodgkin's disease also provides an example of the importance of host features in determining the outcome of therapy. The orderly and predictable patterns of distribution in Hodgkin's disease, which afford hope of cure using radiotherapy even for widespread nodal disease, have given impetus to the recent trend toward more precise staging procedures. The combined application of chemotherapy and radiation therapy, in an effort to improve the results of treatment for patients with poor prognostic features, makes the management of the malignant lymphomas one of the best examples of the effectiveness of a multidisciplinary approach to neoplastic disease.

The leukemias are a group of diseases that have pathogenetic similarities but are widely divergent in clinical manifestations. They share the common feature of being neoplastic proliferations of hematopoietic cell precursors in which maturation and functional capability is impaired. Proliferation is thought to begin in the marrow, where it remains most prominent. Lymph nodes, spleen, and other tissues can also show accumulations of leukemic cells and proliferation in situ.

Routine diagnostic workup for previously untreated patients with lymphomas includes careful history with emphasis on constitutional symptoms, physical examination focused on peripheral lymph nodes, liver, spleen, and abdominal masses. Complete blood count, erythrocyte sedimentation rate, serum alkaline phosphatase, and electrophoresis also are included. Chest x-ray and lower extremity lymphangiography are routinely obtained. CT of the abdomen and bone marrow biopsy are routinely recommended for patients with non-Hodgkin's lymphomas.

Gallium-67 imaging has been useful in the pretreatment detection of occult disease and in the follow-up after institution of treatment for patients with malignant lymphoma or leukemia. In these diseases, the detectability of disease sites by the Ga-67 imaging varies with the tumor histology and the anatomic location. Ga-67 imaging may play a role in the initial staging of lymphoma, although the sensitivity for disease in the abdomen is low. The lymphatic system and regional lymph nodes can also be imaged by radionuclide lymphangiography, a simple, reliable, and reproducible technique for the evaluation of multiple lymph node groups.

Compared to the radiographic bone survey, radionuclide bone scans show earlier and more frequent indications of bone involvement by lymphoma or leukemia, which usually represents a late complication. The accuracy of the In-111 marrow scan in reflecting the degree of normal marrow cellularity is very high, and the marrow scan also assists the clinician in selecting a biopsy site with a high probability of diagnostic yield in patients with lymphoma. Bone marrow scanning may also reveal the extent of active bone marrow regeneration in previously treated patients.

CHARACTERISTICS OF HEMATOLOGIC TUMORS

Epidemiology

Hodgkin's disease constitutes about 40 percent of malignant lymphomas. The incidence rates are 25 cases per million white males and 26 cases per million white females.[1] In 1975, Hodgkin's disease constituted about 1.1 percent of all cancers in the United States.[2] In 1983 it was estimated that 7,100 new cases occurred in the United States.[3] About 50 percent of cases occurred between the ages of 20 and 40. A bimodal occurrence, with peak incidence in the third decade and another high frequency in later life, has been observed. The incidence is slightly higher in males, and males tend to have more extensive disease at diagnosis.

Epidemiologic studies have suggested an increased incidence of Hodgkin's disease in certain groups with moderately close contact,[4] raising the possibility of environmental as well as genetic factors. There is evidence of a cell-mediated immunologic deficiency in patients with Hodgkin's disease,[5] and the demonstration of RNA in Hodgkin's tissue also has strengthened the hypothesis of viral etiology of Hodgkin's disease.[6]

The relative incidence of Hodgkin's to non-Hodgkin's lymphoma is about 2.2 to 1.[7] The peak age incidence of non-Hodgkin's lymphoma is later than for Hodgkin's disease, and about 25 percent of cases develop in the fifth decade. The recent association of large cell lymphoma with chronic immunosuppressive therapy in man seems compatible with either a viral etiology or an induced immunologic defect permitting proliferation of a malignant clone.[8]

In 1975 leukemias made up about 3 percent of neoplasias in the United States.[2] There were an estimated 23,900 new cases of leukemia in the United States in 1983.[3] The incidence rates of leukemias vary according to the type of leukemia, and the frequency of a given type is dependent upon age, sex, race, and locale. Acute leukemia occurs most frequently in the first 5 years of life. It predominates in male infants and is most often lymphatic.[9] Acute myelocytic leukemias are found in male individuals. Chronic myelocytic leukemia is found most frequently in men 20 to 60 years of age, predominantly in those between 30 and 39 years old, and is rarely observed in children.[10] Chronic lymphocytic leukemia is prevalent between 45 and 60 years of age, and about 75 percent of the cases are found in men.

Because ionizing radiation is capable of causing experimental leukemia, it may contribute to the incidence of leukemia in some individuals. However, no significant increase in leukemia has been observed in the offspring of atomic bomb survivors.[11] An increased occurrence of leukemia among shoe workers exposed to benzene has been reported,[12] and there are suggestions that human leukemia like lymphoma may be caused by viruses, and consequently transmissible.[13]

Pathology

A classification of non-Hodgkin's lymphomas is given in Table 1, and the pathology and distribution of a sample of patients is broken down in Table 2.

Accurate distinctions between well and poorly differentiated lymphocytic lymphomas and of large cell lymphomas, nodular or diffuse, will direct staging procedures and provide prognostic indicators.[15]

Attempts have been made to divide Hodgkin's disease into various categories according to the microscopic characteristics. Lukes[16] has proposed a useful classification now generally followed. The approximate occurrence is given in parentheses: lymphocyte predominance (5 percent), nodular sclerosis (52 percent), mixed cellularity (37 percent), and lymphocyte depletion (6 percent).

The majority of cases of Hodgkin's disease present in lymph nodes close to the body's large vessels. The cervical nodes may obstruct veins or invade

TABLE 1. MODIFIED RAPPAPORT CLASSIFICATION OF NON-HODGKIN'S LYMPHOMAS[14]

Nodular	Diffuse
Poorly differentiated lymphocytic	Well differentiated lymphocytic
Mixed lymphocytic–histiocytic	Intermediate differentiated lymphocytic
Large cell (histiocytic)	Poorly differentiated lymphocytic
	Mixed lymphocytic–histiocytic
	Large cell (histiocytic)
	Undifferentiated, Burkitt's
	Undifferentiated, non-Burkitt's
	Lymphoblastic
	Unclassified

TABLE 2. PATHOLOGY AND DISTRIBUTION OF NON-HODGKIN'S LYMPHOMA—405 CASES[15]

Histology	% of Series	% Bone Marrow Involvement	% G.I. Involvement
Lymphocytic, WD			
Nodular	2	40	17
Diffuse	2	43	20
Lymphocytic, PD			
Nodular	17	36	7
Diffuse	11	29	14
Mixed			
Nodular	18	13	4
Diffuse	11	15	20
Histiocytic			
Nodular	7	11	14
Diffuse	29	9	24
Stem Cell			
Diffuse	3	0	43

the muscle. Those of the mediastinum and hilar region are frequently the point of departure for secondary involvement of the trachea, bronchi, pleura, or lungs. The retroperitoneal nodes may involve nerves and the vertebral bodies, and may displace or occlude the ureters. Lung involvement frequently is observed at autopsy, and bone involvement also is found in a large number of cases. At autopsy, the reported occurrence of bone involvement ranges from 10 to 35 percent.[17] The vertebrae, sternum, femoral head, and—rarely—the ribs, pelvis, and skull are involved. The spleen is involved at autopsy in 60 to 70 percent of the cases. It is not usually greatly enlarged, but presents involvement in the form of nodular masses. Involvement of the liver or gastrointestinal tract is found in over 10 percent of the cases. Cerebral involvement is rarely observed.

The pathologic alterations in all types of chronic leukemia tend to involve many systems. The pathologic changes are mainly of two types: those that affect the blood-forming organs, particularly bone marrow, spleen, and lymph nodes, and those due to infiltration. In chronic lymphocytic leukemia, generalized lymphadenopathy is frequent. Moderate generalized lymph node enlargement may occur terminally in myelocytic leukemia. Leukemia infiltration is common in the kidneys, is usually bilateral, and produces enlargement. Leukemic infiltration of the GI tract is also common, particularly in the lymphocytic variety, and is most prominent in the stomach and ileum. Leukemic infiltration of the skin is present particularly from the lymphocytic form of leukemia. There is a greater tendency to lymph node involvement in acute lymphocytic than in the acute myelocytic and monocytic varieties. The liver is invariably enlarged in all cases, and the spleen also is enlarged, but not nearly as much as in the chronic forms of leukemia.

The T and B cell markers have become as important in the classification and understanding of leukemias as they are in the malignant lymphomas. B cells are present in most cases of chronic lymphocytic leukemia, and T cells dominate in occasional cases of chronic lymphocytic leukemia and in some cases of acute lymphocytic leukemia.

Clinical Evolution

The most frequent symptom of Hodgkin's disease is a painless enlargement of lymph nodes. The cervical, axillary, inguinal, and retroperitoneal lymph nodes are most commonly affected. Growth of lymph nodes in various locations may cause backache, dyspnea, edema, or pain of the leg.

Enlargement of the spleen may be noted at some time during the course of the disease, but enlargement of the liver is not usually evident until the terminal stages of disease. Lassitude, weight loss, anemia, and fever occur usually after the disease has become disseminated. In children, fever is a common presenting symptom, and herpes zoster is relatively frequent in patients with Hodgkin's disease.[18]

A generally accepted "Ann Arbor" classification of anatomic staging is important to facilitate meaningful communication, and to aid in planning therapy.[19]

Stage I: Involvement of a single lymph node region (I) or of a single extra-lymphatic organ or site (I_E).

Stage II: Involvement of two or more lymph node regions on the same side of the diaphragm (II) or localized involvement of extra-lymphatic organ or site and one or more lymph node regions on the same side of the diaphragm (II_E).

Stage III: Involvement of lymph node regions on both sides of the diaphragm (III), which may also be accompanied by localized involvement of extra-lymphatic organ or site (III_E), by involvement of the spleen (III_S), or both (III_{SE}).

Stage IV: Diffuse or disseminated involvement of one or more extra-lymphatic organs or tissues with or without associated lymph node enlargement.

Each stage is subdivided into (A) and (B) categories; (B) for those with general symptoms (unexplained weight loss of more than 10 percent of body weight in the 6 months prior to admission, unexplained fever with temperatures above 38°C and night sweats), and (A) for those without such symptoms.

Early involvement of oropharyngeal lymphoid tissue, skin, gastrointestinal tract, and bone is somewhat more common in non-Hodgkin's lymphoma. In children, initial intra-abdominal manifestations occur relatively commonly (over one-third) in non-Hodgkin's lymphoma as contrasted to Hodgkin's disease. Pruritus is not usually present, and fever is rarely observed in early cases of non-Hodgkin's lymphoma.

The onset of acute leukemia is often sudden, but in most patients there is a prodromal period of weakness and malaise followed by fever, tachycardia, and prostation. Fever, pain in the bones and joints, petechia, hemorrhage, and severe secondary infection are frequently cases of acute leukemia.

In chronic lymphocytic leukemia, the outstanding first sign is the enlargement of the lymph nodes, particularly in the cervical region. There is considerable enlargement of the spleen in chronic myelocytic leukemia. In later stages of chronic leukemia, the symptomatology and the clinical finding are protean, with widespread involvement of multiple organs. Anemia, congestive heart failure, and neurologic complications may appear in chronic leukemia, and intense pain may result from splenic infarction.

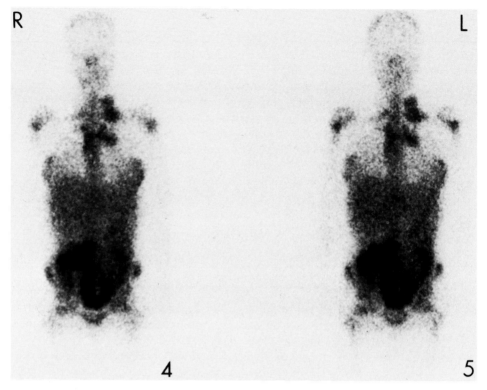

Figure 1. Selective longitudinal tomographic images using Ga-67 citrate show involvement by Hodgkin's disease in the left clavicular and bilateral hilar lymph nodes. Normal uptake of radiogallium in both breasts and normal Ga-67 excretion in the intestine are noted.

NUCLEAR IMAGING

Ga-67 Imaging

While precise mechanisms of gallium localization in neoplastic processes are unknown, there is a greater body of empirical data concerning applications to various disease entities. One of the first applications of gallium scan was for the staging of malignant lymphomas.[20] Ga-67 scans (Fig. 1) have proved useful in the detection of sites of lymphoma involvement above the diaphragm, but have had a low diagnostic accuracy below the diaphragm.[21] Lesions in or near the liver, spleen, and iliac node regions are difficult to detect due to overlying organ activity. With a high resolution imaging device and a high dose (10 mCi) of Ga-67, the sensitivity of the scan for intraabdominal disease sites (Fig. 2) (except the liver and spleen) is at least 0.50 with a specificity of over 0.95.[22] Ga-67 scanning is probably as good as radiography for detecting disease in the mediastinum (Fig. 3). Its sensitivity for the mediastinal lesion may appear higher than in fact it is, since small lesions may be undetected by any diagnostic procedure.[23] Horn et al.[21] found that the sensitivity of Ga-67 scanning is significantly greater for sites of Hodgkin's disease or histiocytic lymphoma than sites of lymphocytic or mixed lymphocytic histiocytic lymphoma.

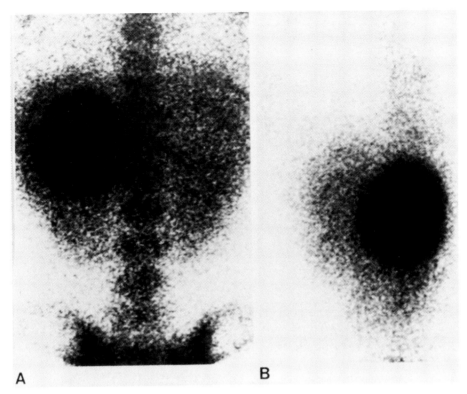

Figure 2. Posterior and left lateral views of the abdomen 48 hours following Ga-67 citrate injection show a large area of abnormal uptake of radiogallium in the left upper abdomen. Biopsy showed a malignant fibrous histiocytoma. *(From Am J Roentgenol 1980; 135:776, with permission.)*

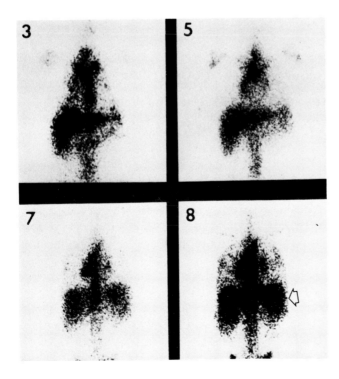

Figure 3. Selective longitudinal tomographic images of chest and abdomen with Ga-67 citrate show involvement by Hodgkin's disease in hilar nodes as well as mediastinum. Also noted is splenomegaly due to Hodgkin's disease confirmed at surgery (arrow).

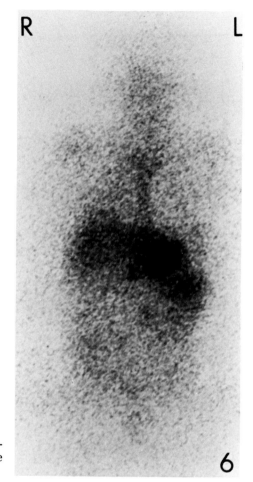

Figure 4. Longitudinal section image using Ga-67 citrate shows abnormal uptake of radiogallium in a gastric lymphoma.

A diagnosis of lymphoma cannot be made by Ga-67 scanning alone due to its nonspecificity. However, Ga-67 scanning may influence the initial staging and management of patients with malignant lymphoma under appropriate circumstances. Many false negative results were found in the cervical, axillary, and inguinal areas, whereas the majority of true positive scans were found in the mediastinum and also in the lung parenchyma.[24] Extensive gastric lymphomas (Figs. 4 and 5) were better demonstrated by the Ga-67 imaging than by the upper GI series.[25] In those cases of malignant lymphoma in which exploratory laporotomy is not indicated, or in situations in which patients refuse more conventional examinations, the Ga-67 scan may be used for staging.

Leukemia is a special case in which diffusely increased Ga-67 uptake throughout the skeleton may be indicative of disease. This pattern is seen in more than 75 percent of patients with acute leukemia and extensive marrow infiltration. Similar findings have also been observed in chronic myelogenous leukemia with blast crisis. Leukemic infiltrates such as myeloblastomas may localize gallium.

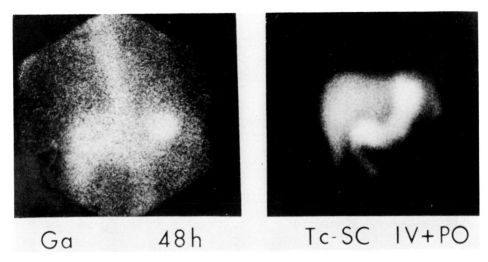

Ga   48h   Tc-SC   IV+PO

Figure 5. Anterior image of chest and upper abdomen (left) 48 hours following Ga-67 citrate injection shows focal uptake in a gastric lymphoma that was localized after giving Tc-99m sulfur colloid orally (right).

Ga-67 scan (Fig. 6) revealed leukemic infiltration of such organs as the heart or testis, while CT and ultrasonography failed to demonstrate evidence of occult disease in these organs.[26,27] Ga-67 imaging appears more useful for reevaluating patients after therapy because it is the simplest and least invasive way of doing so.

Radionuclide Liver-Spleen Imaging

In most patients, Hodgkin's disease invades the liver through the portal venous system. After initial spread through contiguous lymph node groups, Hodgkin's disease may involve the splenic parenchyma (Fig. 7). If the spleen is not palpable, there is approximately a 5 percent incidence of liver involvement, and a palpable spleen (Fig. 8) increases the chance of liver involvement tenfold. In spleens greater than 400 g there is an 81 percent incidence of hepatic tumor.[28] Fialk et al.[29] reported a patient with Hodgkin's disease who at laparotomy had liver involvement with no evidence of Hodgkin's disease in the spleen. The increased lymphatic pressures in the presence of massive retroperitoneal lymphadenopathy may result in the formation of lymphaticovenous anastomoses, thus providing a possible second route for invasion of the portal venous system and subsequently the hepatic parenchyma.[29]

The published data concerning the value of radionuclide liver-spleen scanning methods in the investigation of hematologic malignancy is inconclusive. Jasinski and Mikolajakow[30] considered that the uptake pattern of Tc-99m sulfur colloid correlated well with the clinical course of the disease. A patchy distribution of the radioactivity or filling defects may be observed in patients with Hodgkin's disease or histiocytic lymphoma (Figs. 9 and 10). Spleens weighing more than 400 g are found occasionally to have no pathological involvement. Conversely, a normal-sized spleen with a normal uptake pattern

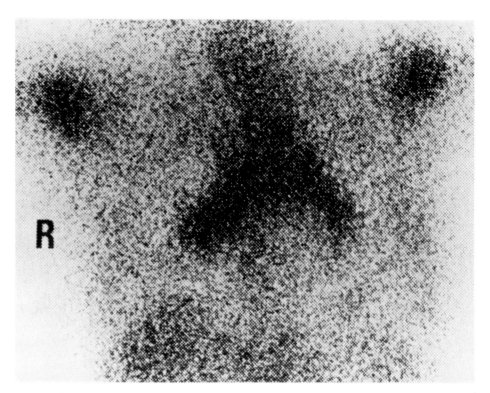

Figure 6. Anterior chest image with Ga-67 citrate shows an uptake of radiogallium in the thymus of a 12-year-old male with T-cell leukemia and mediastinal widening by chest x-ray. *(From J Nucl Med 1981; 22:1043, with permission.)*

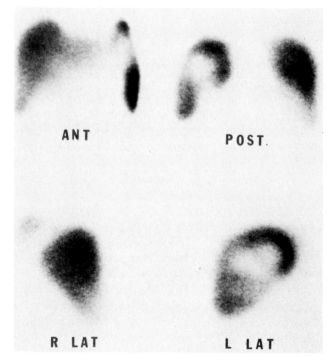

Figure 7. Routine liver-spleen images with Tc-99m sulfur colloid show a large defect in the spleen due to Hodgkin's disease.

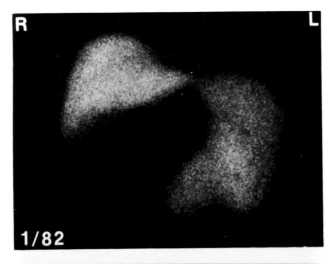

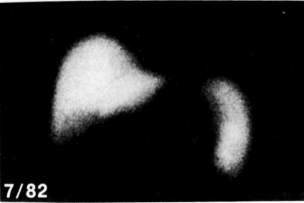

Figure 8. Anterior liver-spleen image with Tc-99m sulfur colloid shows marked splenomegaly due to non-Hodgkin's lymphoma involvement. Follow up study 6 months after chemotherapy showed marked reduction of splenic size.

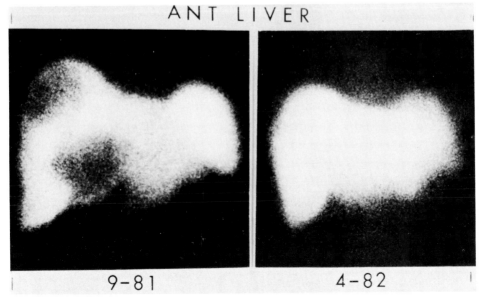

Figure 9. Anterior colloid liver-spleen image shows large defects in the right lobe of liver due to Hodgkin's disease. Follow-up study 7 months after chemotherapy showed clearing of these lesions.

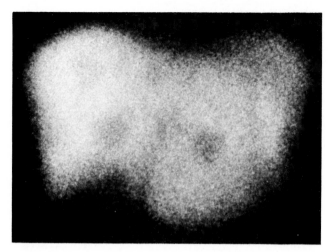

Figure 10. Anterior colloid liver-spleen image showed multiple focal defects in an enlarged liver with a prominent left lobe. Autopsy revealed involvement by Kaposi's sarcoma.

does not exclude Hodgkin's disease.[31] Leukemic involvement usually demonstrates mild to moderate splenomegaly (Fig. 11).

Radionuclide Bone Imaging

Bone involvement (Fig. 12) is a late development in Hodgkin's disease. Ultmann,[32] who used the skeletal survey for assessment of bone pathology, found lesions in 6 percent of patients at the time of initial staging. Fifteen to 20 percent of patients developed radiographic evidence of bone involvement

Figure 11. Anterior image of liver-spleen using Tc-99m sulfur colloid showed marked splenomegaly due to involvement by chronic myelocytic leukemia.

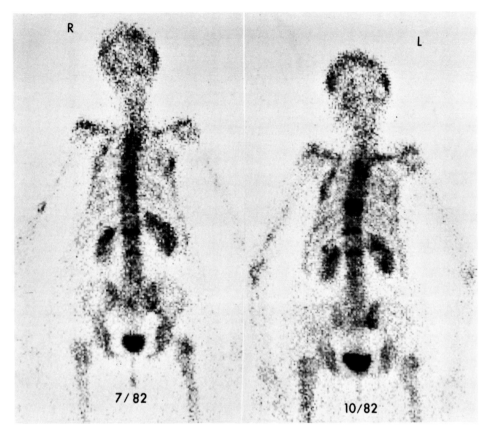

Figure 12. Posterior images of axial skeleton with Tc-99m MDP show multiple areas of increased activity in the spine, scapulae, left pelvis, right humerus, and left femur due to Hodgkin's disease. Note mixed responses following chemotherapy in 3 months.

during the course of their disease. Radionuclide bone scans (Fig. 13) appeared to give earlier indication of bone involvement than the radiographic bone survey in 51 patients with Hodgkin's disease.[33] In 33 patients with chronic myelogenous leukemia, the bone scans (Fig. 14) were abnormal in 67 percent and the radiographs in 29 percent.[34] Because of the lower yield of positive findings in asymptomatic patients, bone scans (Fig. 15) have been indicated only in patients with bone pain.

Radionuclide Bone Marrow Imaging

The evaluation of marrow infiltration with lymphoma frequently involves the histologic examination of marrow obtained by a marrow biopsy or aspiration. Nonuniformity in post-treatment regeneration of marrow and focal tumor involvement of the marrow are conditions that tend to increase the likelihood that a random marrow biopsy will not be representative of the marrow organ. Marrow scan may also be helpful to examine the extent of active bone marrow regeneration in previously treated patients.

The In-111 marrow scan reflects the degree of normal marrow cellularity

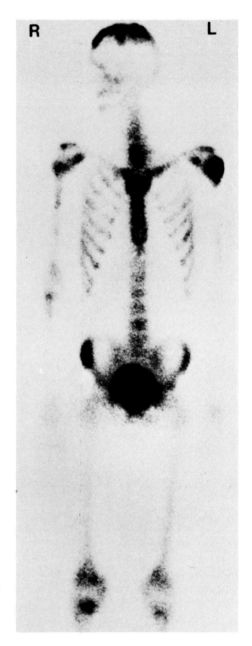

Figure 13. Anterior image of whole-body scan with Tc-99m MDP shows multiple areas of lymphoma involvement in skull, humeri, and right iliac crest.

in each site subjected to biopsy with an accuracy of 90 percent.[35] One hundred and two previously treated lymphoma patients were studied with In-111 scans, and the accuracy of the scan was confirmed by the marrow biopsy or the subsequent clinical course of the patient.[36] Fibrosis or infiltration of the marrow with large amounts of Hodgkin's disease (Fig. 16) was associated with decreased uptake of the radiopharmaceutical.

Chemotherapy causes generalized depression of uptake, and extended irradiation also depresses the uptake within the irradiated field. Acute

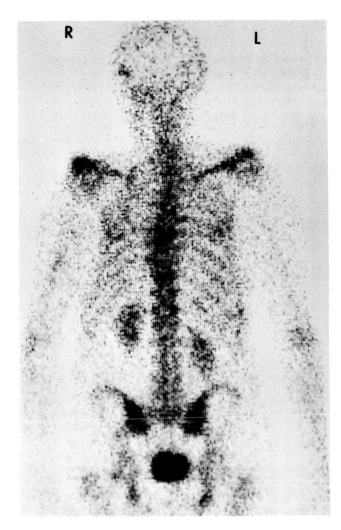

R   L

Figure 14. Posterior body scan with Tc-99m MDP showed slightly increased activities in both pedicles of T-7 and T-10 vertebrae, and left pedicles of T-11 and T-12 vertebrae, due to involvement by chronic myelogenous leukemia.

leukemia demonstrates less accumulation of In-111 chloride than normal and chronic leukemia revealed widespread distribution of the radioactivity.[37]

Radionuclide Lymphangiography

Lymphoscintigraphy is based on the mechanism of the transport of a radioactive colloid injected into the subcutaneous tissue. While the initial studies with this technique were performed in the ileopelvic lymph node groups, lymphoscintigraphy has been more widely used in the internal mammary lymphatics in patients with carcinoma of the breast. Ileopelvic lymphoscintigraphy, when performed following bilateral perianal injections of Tc-99m antimonytrisulfide colloid, offers a physiologic method for visualizing these nodes and allows demonstration of the internal iliac chain, often not visualized on contrast lymphangiographic studies.[38] Compared with the ethiodol lymphangiographic results positive lymph scans revealed an accuracy rate of 83 percent[39] in the group of patients with lymphoma and Hodgkin's disease.

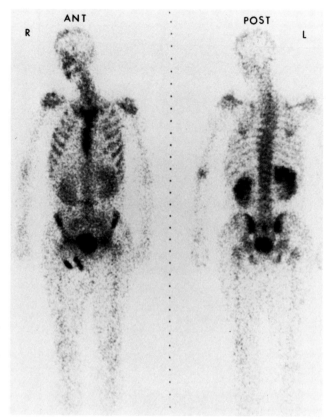

Figure 15. Anterior and posterior whole-body images with Tc-99m MDP showed normal bone structure but significantly increased activity in the upper portion of left kidney. Surgery confirmed lymphoma.

The identification of abnormalities within the lymph nodes draining a particular tumor may suggest aggressive treatment programs that may result in an improved potential for survival. This noninvasive technique of lymph node visualization also allows repetitive studies in follow-up offering earlier evidence of recurrent disease than can be achieved by many of the diagnostic techniques, which are sometimes associated with a high false-negative rate.

Other Nuclear Studies

In normal subjects the lymphocytes leave the blood initially, but later return to it due to the reappearance of cells that at first entered the spleen. Using In-111 oxine labeled lymphocytes, it was found in patients with chronic lymphocytic leukemia that no such reappearance in the blood occurs, and the lymphocytes do not leave the spleen in significant numbers over 48 hours.[40] Dynamic studies of In-111-labeled lymphocytes were also reported after treatment of T-cell lymphoma patients with hybridoma monoclonal antibody.[41] The results suggest that the hepatic reticuloendothelial system is involved in the removal of lymphocytes, and that anti-tumor antibodies have therapeutic potential.

Despite extensive lymphoma involvement, the Tl-201 myocardial scan was normal in one patient, probably related to uptake of Tl-201 by lymphoma cells.[42] Spinal meningeal uptake of Tc-99m MDP in meningeal seeding by malignant lymphoma has been reported.[43]

Patients with non-Hodgkin's lymphoma and splenomegaly have a high

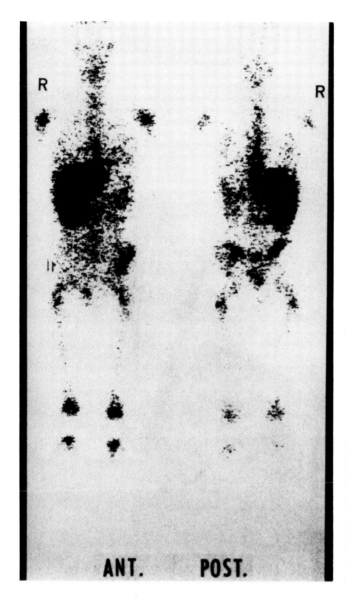

R

R

ANT.   POST.

Figure 16. Anterior and posterior whole-body images of bone marrow scan using In-111 chloride show decreased uptake in the right anterior iliac crest. Biopsy revealed total replacement with Hodgkin's disease. *(From Cancer 1976; 38:1566, with permission.)*

incidence of deep vein thrombosis (DVT) after splenectomy. However, there was no significant excess of DVT as measured by the I-125 fibrinogen uptake in patients having staging laparotomy and splenectomy for Hodgkin's disease.[44]

OTHER RADIOLOGIC EXAMINATIONS

Radiographs
The chest x-ray should be performed in every patient suspected of having lymphoma or leukemia, because of the frequent involvement of the mediastinum and lung parenchyma. Positive findings will often be revealed in spite of the absence of clinical symptoms.

Full chest tomograms may prove helpful in demonstrating extent of mediastinal lymphadenopathy or pulmonary involvement not revealed in ordinary radiographs. However, the yield of whole-lung tomography in patients with a negative conventional chest x-ray is too low to justify its use as a routine procedure.

The higher incidence of bone involvement by Hodgkin's disease reported from autopsies indicates that many bone lesions escape detection by conventional radiographic examination. When the involvement occurs from contiguous lymph nodes, destruction of the cortex will be demonstrated easily. Bone lesions are usually of a mixed osteoblastic and osteolytic type. Lesions of the skull are invariably osteolytic. Bone changes by leukemia frequently are observed in children; they are usually osteolytic, but can be mixed in character. Destructive lesions of bone in the adult with leukemia are uncommon, and the appearance of bony lesions may accompany the blastic transformation in chronic myelocytic leukemia.

Contrast Lymphangiography, Urogram, and GI Studies

Contrast lymphangiography demonstrates retroperitoneal disease in 80 to 90 percent of cases considered to represent clinical stage I undifferentiated Hodgkin's lymphoma as compared to 30 percent in clinical stage I Hodgkin's disease.[45] Awareness of this tendency to early widespread disease is important in planning therapy. Negative lymphangiograms have been correlated with low incidence of bone marrow, liver, and spleen involvement. Complications, such as oil embolism, should be kept in mind, especially in older patients with probable generalized disease and/or pulmonary disease. The discomfort and cost of lymphangiograms are also significant factors. Ureteral deviation on IVP generally may eliminate the need for lymphangiograms. Upper and lower gastrointestinal tract barium studies are indicated only in patients with gastrointestinal complaints, since the yield of positive findings is negligible in asymptomatic patients.

CT and Ultrasonography

CT of the thorax or abdomen or ultrasonography of the abdomen have been useful when lymphangiography is medically contraindicated, technically unsatisfactory, or results in incomplete filling of the upper lumbar retroperitoneal lymph nodes. The accuracy of CT scans[46] was virtually identical to lymphangiography for the detection of para-aortic lymph node involvement; however, CT provides a better means of assessing the true extent of the disease. Moreover, this excellent definition permits accurate follow-up assessment of therapy. The CT or ultrasonography is also an excellent means of guiding biopsy procedures even for retroperitoneal abnormalities, perhaps precluding the necessity for laparotomy in order to provide histologic diagnosis of disease.

REFERENCES

1. MacMahon B: Epidemiology of Hodgkin's disease. Cancer Res 1966; 26:1189–1200

2. Cutler SJ, Young JL, Jr: Third national cancer survey: Incidence data. Natl Cancer Inst Mongr 1975; 41:10–27, 100–135, 388–427
3. Silverberg E: Cancer statistics, 1983. Ca-A Cancer J Clinicians 1983; 33:9–25
4. Vianna NJ, Greenwald P. Brady J, et al: Hodgkin's disease: Cases with features of a community outbreak. Ann Int Med 1972; 77:169–180
5. Cole P: Epidemiology of Hodgkin's disease. JAMA 1972; 222:1636–1639
6. Spiegelman S, Kufe D, Hehlmann R, et al: Evidence for RNA tumor viruses in human lymphomas including Burkitt's disease. Cancer Res 1973; 33:1515–1526
7. Rosenberg SA, Diamond HD, Jaslowitz B, et al: Lymphosarcoma: A review of 1269 cases. Medicine 1961; 40:31–36
8. Penn I: Malignant tumors arising de novo in immunosuppressive organ transplant recipients. Transplant 1972; 14:407–417
9. Cutler SJ, Axtell L, Heise H: Ten thousand cases of leukemia: 1940–1962. J Natl Cancer Instit 1967; 39:933–943
10. Kantha KRR, King M, Levy RN, et al: Chronic myelogenous leukemia in childhood. NY State J Med 1975; 75:392–399
11. Hoshino T, Kato H, Finch SC, et al: Leukemia in offspring of atomic bomb survivors. Blood 1967; 30:719–730
12. Aksoy M, Erdem S, Dincol G: Leukemia in shoe-workers exposed chronically to benzene. Blood 1974; 44:837–841
13. Bryan WR, Moloney JB, O'Connor TE, et al: Viral etiology of leukemia. Ann Intern Med 1965; 62:376–399
14. Callihan TR, Berard CW: The classification and pathology of the lymphomas and leukemias. Semin Roentgenol 1980; 15:203–218
15. Jones SE, Rosenberg SA, Kaplan HA, et al: Non-Hodgkin's lymphoma. Clinico-pathologic correlation in 405 cases. Cancer 1973; 31:806–823
16. Lukes RJ: The pathologic picture of the malignant lymphomas, in Zarafonetis CJD (ed): Proceedings of the International Conference on Leukemia-Lymphoma. Philadelphia, Lea & Febiger, 1968
17. Westling P: Studies of the prognosis in Hodgkin's disease. Acta Radiol 1965; 245:5–124
18. Editorial: Zoster and Hodgkin's disease. Br Med J 1972; 15:130–131
19. Stutzman L: Current concepts in treatment of lymphomas with special reference to Hodgkins' disease. Ann Rev Med 1973; 24:325–334
20. Turner DA, Pinsky SM, Gottschalk A, et al: The use of ^{67}Ga scanning in the staging of Hodgkin's disease. Radiol 1972; 103:97–102
21. Horn, NL, Ray GR, Kriss JP: Gallium-67 citrate scanning in Hodgkin's disease and non-Hodgkin's lymphoma. Cancer 1976; 37:250–257
22. Turner DA, Fordham EW, Ali A, et al: Gallium-67 imaging in the management of Hodgkin's disease and other malignant lymphomas. Semin Nucl Med 1978; 8:205–218
23. McCaffrey JA, Rudders RA, Kahn PC, et al: Clinical usefulness of Ga-67 scanning in the malignant lymphoma. Am J Med 1976; 60:523–528
24. Huys J, Schelstraete K, Simons M: Ga-67 imaging in Hodgkin's disease. Clin Nucl Med 1982; 7:174–179
25. Ichiya Y, Oshium Y, Kamoi I, et al: Ga-67 scanning and upper gastrointestinal series for gastric lymphomas. Radiol 1982; 142:187–192
26. Milder MS, Glick JH, Henderson ES, et al: [67]Ga scintigraphy in acute leukemia Cancer 1973; 32:803–808
27. Mahoney DH, Gonzales ET, Ferry GD, et al: Childhood acute leukemia. A search for occult extramedullary disease prior to discontinuation of chemotherapy. Cancer 1981; 8:1964–1966
28. Rosenberg, SA: Annotation. Br J Haematol 1972; 23:271–276

29. Fialk MA, Jarowski CI, Coleman M, et al: Hepatic Hodgkin's disease without involvement of the spleen. Cancer 1979; 43:1146–1147

30. Jasinski WK, Mikolajakow A: Clinical significance of the liver and spleen uptake of colloidal Tc-99m, in Medical Radioisotope Scintigraphy, Vienna, IAEA, 1973, vol 2, pp 29–44

31. Ell PJ, Britton KE, Farrer-Brown G, et al: An assessment of the value of spleen scanning in the staging of Hodgkin's disease. Br J Radiol 1975; 48:590–593

32. Ultmann JE: Clinical features and diagnosis of Hodgkin's disease. Cancer 1966; 19:297–307

33. Harbert JC, Ashburn WL: Radiostrontium bone scanning in Hodgkin's disease. Cancer 1968; 22:58–63

34. Valimaki M, Vuopio P, Liewendahl K: Bone lesions in chronic myelogenous leukemia. Acta Med Scand 1981; 210:403–408

35. Gilbert EH, Goris ML, Earle JD, et al: The accuracy of ^{111}InCl$_3$ as a bone marrow scanning agent. Radiol 1976; 119:167–168

36. Gilbert EH, Earle JD, Glastein E, et al: ^{111}Indium bone marrow scintigraphy as an aid in selecting marrow biopsy sites for the evaluation of bone marrow elements in patients with lymphoma. Cancer 1976; 38:1560–1567

37. Ino T, Takeuchi A, Kawai Y, et al: Clinical study of bone marrow scintigraphy with In-111 chloride. Radioisotopes 1980; 29:614–617

38. Kaplan WD: Iliopelvic lymphoscintigraphy. Semin Nucl Med 1983; 13:42–53

39. Croll MN, Brady LW, Dadparvar S: Implications of lymphoscintigraphy in oncologic practice: Principles and differences vis-a-vis other imaging modalities. Semin Nucl Med 1983; 13:4–8

40. Wagstaff J, Gibson C, Thatcher N, et al: The migratory properties of In-111 oxine labeled lymphocytes in patients with chronic lymphocytic leukemia. Br J Haematol 1981; 49:283–291

41. McDougall IR, Goris ML, Miller RA: Dynamic studies of In-111 labeled lymphocytes after treatment with hybridoma monoclonal antibody. Clin Nucl Med 1981; 6:452

42. McDonnell PJ, Becker LC, Bulkley BH: Thallium imaging in cardiac lymphoma. Am Heart J 1981; 101:809–814

43. Siegel T. Or R, Matzner Y, et al: Spinal meningeal uptake of Tc-99m MDP in meningeal seeding by malignant lymphoma. Cancer 1980; 46:2413–2415

44. Dawson AA, Bennett B, Jones PF, et al: Thrombotic risks of staging laparotomy with splenectomy in Hodgkin's disease. Br J Surg 1981; 68:842–845

45. Keller AR, Kaplan HS, Lukes RJ, et al: Correlation of histopathology with other prognostic indicators in Hodgkin's disease. Cancer 1968; 22:487–492

46. Zelch MG, Haaga JR: Clinical comparison of computed tomography and lymphangiography for detection of retroperitoneal lymphadenopathy. Radiol Clin N Am 1979; 17:157–168

CHAPTER 12

New Modalities for Nuclear Imaging

RADIOIMMUNODETECTION OF CANCER

Most radiopharmaceuticals for tumor imaging are nonspecific and limited-use agents. With these agents, abnormalities on scan represent alteration or displacement of normal tissue. Hence, space-occupying lesions may be due to non-neoplastic processes and other measures such as a biopsy may be necessary to accurately define the etiology of the abnormality detected by scan. Additional scans with other radiopharmaceuticals are often necessary to evaluate potential involvement of other organs because of limited-use radiopharmaceuticals. They would include I-131 for well differentiated thyroid cancer, I-131 iodocholesterol for functioning adrenal cortical tumors, and radiolabeled antibodies to tumor-associated antigens. They are relatively specific for a particular organ or type of tissue, and they allow distinction between tumor and normal tissue with external imaging because of some special property of the tumor.

The immunologic approach to tumor detection utilizing antigen-antibody reactions appears to offer a specific method based on the thesis that tumors may bear unique tumor-specific antigens. The antibodies localize and specifically concentrate in each tumor area providing guidance for surgical, radiotherapeutic, or chemotherapeutic intervention. However, almost all tumor antigens described to date have been shown to be only quantitatively increased in certain tumors.[1] Tumor-associated markers (e.g., oncofetal protein antigens, enzymes, hormones, or polyamines) may be quantitated in the serum and other body fluids. In spite of nonspecificity and association with non-neoplastic disease, they may be used in attempting to detect and treat a malignant tumor before it metastasizes. It should be noted, however, that changing levels of the tumor marker may not always mirror a change in the tumor burden. With treatment of the tumor, a decrease in the quantity of a tumor-marker may signify that there has been a response to the therapy and relapse of a tumor is usually associated with the reappearance of the marker, but the marker may not be present with recurrence since some tumors lose their ability to produce the marker.

Recently, Pressman[2] reviewed historical aspects of the development of radiolabeled antibodies for tumor localization. Following the development of a kidney antibody labeled with I-131 in 1948, several studies reported localization of transplanted tumors in rodents, using radiolabeled antibodies to fibrin. In the study of 50 patients with various types of malignant tumors, immuno-scintiscanning with I-131-labeled rabbit antibodies to human fibrinogen had an overall accuracy of tumor localization of 58 percent,[3] and the best results were obtained in localization of brain metastases. Two patients were given therapeutic doses (100 and 160 mCi:2000 rad calculated tumor dose) with I-131 fibrinogen antibody, and the tumor in one of them was noted to have decreased to 30 percent of its former size as estimated by physical examination and radiographic findings.

There are a number of characteristics of an antigen that are necessary to obtain a meaningful radiolabeled, localizing antibody. It appears important that the antigen be located on the surface membrane of the tumor cell in order to react with antibody, as it has not been shown that antibodies are able to penetrate the membrane of living cells. It also appears important that such membrane antigens have a reasonably long half-life on the surface of the cell, because the antigens may be released from the tumor to the general circulation where they can combine with parental antibody in vivo and be phagocytized by the liver and spleen. This removes the labeled antibody from circulation and prevents redeposition in the tumor site.[4]

Iodination of the antiserum is relatively simple, with a variety of procedures available. To prevent a decrease or loss of biological specificity and in vivo solubility of the antiserum, no more than two iodines should be attached to the protein molecule.[5] Of particular importance to in vivo studies is the potential for a rapid loss of a significant fraction of a labeled preparation. This results in a very rapid initial decrease in the circulating radioactivity and a high initial uptake in the liver, indicating denaturation of the protein. Another problem involves preparations with a large amount of the isotope attached to material that is not tumor-specific. To reduce this possibility, a purfication procedure should be utilized. The most refined technique is to remove the nonreacting protein before the antibody is labeled with the isotope; however, the difficulty here is ending up with a small amount of protein that is relatively easy to denature.

In the absence of more tumor-specific antigens, much interest has focused on the cell-surface glycoprotein, carcinoembryonic antigen (CEA), which is one of the best characterized tumor-associated oncofetal antigens. Very sensitive radioimmunoassays for CEA are currently used in clinical practice to monitor disease activity in proven cancer patients. Goldenberg et al.[6] have reported gratifying results of radioimmunodetection in 142 patients with a proven history of cancer. They have used 0.7 to 2.1 mCi/70 kg patient (2 to 3 ug/kg immunoglobulin G) I-131 labeled affinity-purified goat immunoglobulin G having 70 percent immunoreactivity with CEA, and computer-

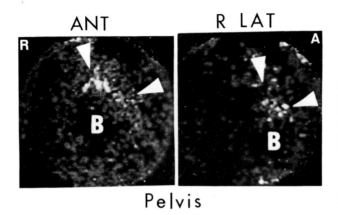

ANT **R LAT**

Pelvis

Figure 1. 24-hour anterior and right lateral pelvic images of (I-Tc) subtraction scan using I-131 anti-hCG IgG and Tc-99m pertechnetate show a localization of hCG-secreting hydatidiform mole (arrows). B indicates urinary bladder.

assisted imagings to subtract Tc-99m background radioactivity for the enhancement of the tumor-related activity. The overall sensitivity in the four major cancer type studies was as follows: colorectal, 85 percent; ovarian cancer, 88 percent; cervical cancer, 90 percent; and lung cancer, 71 percent. The smallest tumors detectable by this method appeared to be about 2 cm in diameter. Metastatic tumors could be localized in a number of patients with normal plasma CEA titers, and large amounts of circulating CEA did not appear to prevent successful localization of tumors. Only two of 116 nonneoplastic benign diseases showed some evidence of radioantibody localization, and only four of 32 tumor sites revealed loss of uptake of I-131 normal goat immunoglobulin G after initially being positive; three of these four sites were at least 10 cm in diameter.

Dykes et al.[7] also has reported similarly good results with imaging of gastrointestinal tract tumors with sheep IgC antibody to CEA. In 13 patients with tumors, 4 out of 5 primary sites and 8 out of 11 secondary sites were successfully localized. Kim et al.[8] has studied 17 patients with malignant tumors by radioimmunodetection, using goat anti-alpha-fetoprotein (AFP) antibody labeled with I-131 and computer-assisted subtraction of Tc-99m background radioactivity. All 12 sites involved by five AFP-producing tumors could be demonstrated by this method, while five patients in whom tumors were not expected to produce AFP also showed positive immunoscintigraphy in six of 16 tumor sites. McManus et al.[9] used peroxidase-labeled antibody against the beta subunit of human chorionic gonadotropin (hCG) to demonstrate that 25 human tumors contained hCG, which was found in the cytoplasm and on the surfaces of the malignant cells. Injection of I-131 labeled goat immunoglobulin G antibody to hCG into patients with hCG-secreting trophoblastic (Fig. 1) and germinal tumors permitted tumor detection and location by external scintigraphy.[10] Excision of one of the metastatic tumors located by this method revealed a tumor to non-tumor radioactivity ratio of 39.29.:1. Hybridomas have recently been used to develop monoclonal tumor-specific antibodies that are homogenous and reproducibly prepared. Suppression of nonspecific background activity may be accomplished by simultaneous administration of a monoclonal nontumor-specific antibody of the same class as the tumor-specific antibody, but labeled with a different radioisotope of the same element. Ballow et al.[11] demonstrated the feasibility of such a method. Mice bearing transplanted teratocarcinoma in one thigh and myeloma in the other thigh were treated with a mixture of I-131 teratocarcinoma-specific monoclonal antibody and I-123 nonspecific antibody. After the radioactivity caused by the indifferent antibody was subtracted from the image of the tumor-specific antibody, localization of residual radioactivity within the thigh teratocarcinoma was evident. Tumor to muscle and tumor to blood ratios of 150:1 and 15:1, respectively, were observed at 5 days after administration of 15 μCi of I-131 labeled hybridoma derived monoclonal antibody to teratocarcinoma-bearing mouse.[12]

Since the Fc portion of immunoglobulins is responsible for much of their nonspecific binding to liver and other tissues, the use of Fab fragments separated from the Fc portion of the native IgG should be advantageous for tumor localization. In addition, the molecular weight of Fab fragments (50,000 daltons) is below the threshold for glomerular filtration, thus allowing for the rapid elimination of unbound, labeled Fab from the organism via the urine.[13] Wilbanks et al.[14] have reported successful localization of mammary tumors (Fig. 2) using the Fab fragments of antibodies against cell-specific surface antigens of MME-antigens, and normalization to Tc-99m pertechnetate distribution in the animal increased the specificity. The potential of this area of research is only just beginning to be realized and all of the present results indicate a significant future for immunoglobulin application in cancer diagnosis and treatment.

RADIOLABELED ANTITUMOR AGENTS

Extensive effort has been directed toward the development of specific, general use radiopharmaceuticals, which could be used for a broad spectrum of tumor types. Radiolabeled antitumor agents are drugs that concentrate in the tumors against which they act, a prerequisite for any imaging agent. The bleomycins are a group of related glycopeptide antitumor antibodies that produced good therapeutic results in squamous cell tumors, lymphomas, sarcomas, and testicular tumors. Tumor imaging studies have been undertaken using bleomycin labeled to Co-57 and In-111. Considering the physical half-life, bleomycin labeled with In-111 (2 to 8 days) would be preferable to Co-57 (120 days). However, a comparative study of Co-57 bleomycin with In-111 bleomycin suggested better images and higher sensitivities with Co-57 bleomycin.[15] A comparison in 50 patients with a broad spectrum of tumors showed that 84 percent overall sensitivity was found for epidermoid carcinoma. In their series, Co-57 bleomycin was also shown to concentrate in suppurative surgical incisions, similar to Ga-67 citrate.

Lilien et al.[16] have evaluated In-111 bleomycin in 293 patients. Among 246 of those with cancer (89 percent) were true positive and 28 (11 percent) false positive. The true positive rates they report are: gastrointestinal 95 percent; lymphoma, 88 percent; melanoma, 75 percent; sarcomas, 82 percent; lung, 77 percent; breast, 77 percent; childhood tumors, 71 percent; gynecologic, 70 percent; and genitourinary tumors, 68 percent. Hepatic and bone marrow uptake of In-111 bleomycin was much greater than with Co-57 bleomycin, suggesting that In-111 either alters the bleomycin molecule or

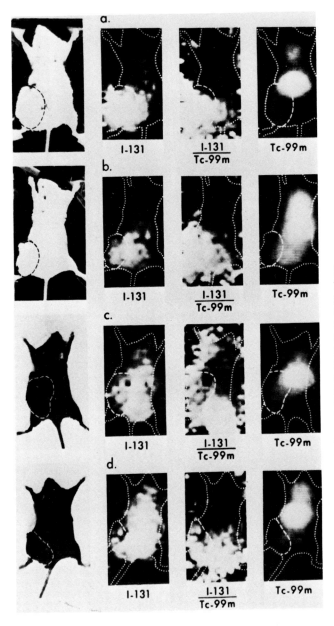

Figure 2. Images obtained on the high purity germanium camera of mice carrying simulated metastases of mammary (**a** and **b**) and nonmammary (**c** and **d**) tumors in the left thighs injected with I-131 anti-MME (Fab) (24 hours before imaging) and Tc-99m pertechnetate (1 hour before imaging). Central image (I-131/Tc-99m) for each mouse is the ratio of I-131 to Tc-99m. (**a**) Mouse mammary tumor (3910-30) in BALB/c mouse; (**b**) mouse mammary tumor (HOG) in BALB/c mouse; (**c**) melanoma (B-16) in C57/BL mouse; (**d**) Lewis lung carcinoma in C57/BL mouse. *(From Cancer 1981; 48:1768, with permission.)*

that it forms a much weaker chelate with bleomycin than does cobalt and free In-111 is released in vivo. With the long physical half-life of Co-57 (270 days) posing serious contamination problems, and the metal chelates of bleomycin having instability in vivo, blocking the native metal-binding chelating group (EDTA) to a nonessential site in the terminal amine region of bleomycin has been tried. The resulting bleomycin conjugate, BLEDTA, when labeled with In-111 had excellent tumor imaging properties (Fig. 3), and a positive scan was obtained in 71 of 88 patients with a positive biopsy (81 percent sensitivity).[17]

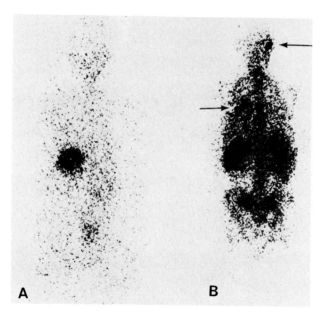

Figure 3. 24-hour posterior whole-body images with I-131 (**a**) and In-111 BLEDTA (**b**) show metastatic lesions of anaplastic thyroid carcinoma in right face and left lung (arrows) localized with BLEDTA *(From J Nucl Med 1981; 22:791, with permission.)*

Mori et al[18] studied 390 patients with Tc-99m bleomycin scanning (Fig. 4) and the overall positive rate was 88 percent compared to 65 percent with Ga-67 citrate. However, technetium appears to be released from the complex and bound to serum albumin in vivo, resulting in delayed clearance from blood.

Unlabeled tetracyclines have been used to demonstrate tumors by their fluorescence under ultraviolet light. However, the concentration of tetracycline in tumor is very low. Tetracyclines labeled to Tc-99m or I-131 have been shown to concentrate in tumors[19]; however, at present there does not seem to be any particular interest in, or advantage from, the use of these compounds.

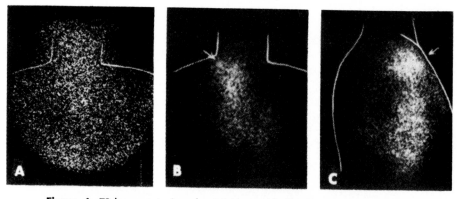

Figure 4. 72-hour anterior chest image with Ga-67 citrate (a), frontal (b), and right lateral (c) views of the neck and chest 60 minutes after Tc-99m bleomycin injection show an accumulation of bleomycin in squamous cell carcinoma in right lower neck (arrows). *(From J Nucl Med 1975; 16:419, with permission.)*

The metabolism of 5-fluorouracil in cancer patients suggests the use of F-18 labeled 5-fluorouracil as a probe for understanding individual response variability in the clinical situation.[20]

TUMOR IMAGING WITH RADIOLABELED AGENTS

Rare earth elements of high atomic number (radiolanthanides) have been found to concentrate within tumors. Yb-169 citrate was used as the tumor scanning agent in 360 patients with malignant tumors and 55 others with benign lesions.[21] In malignant tumors, the overall positive detection rate was 65.3 percent; in benign lesions the false positive rate was 29.1 percent. Good results were obtained in the extremities, head, neck, and pelvic area. Of all tumors visualized, squamous cell carcinoma was detected most effectively. Radiolanthanides are similar to Ga-67 citrate in tissue and cellular distribution. The antitumor activity of various group VIII metal salts has been studied, and in order of antitumor activity against W256 carcinosarcoma was identical to that seen in tissue culture, namely $Tl > Al > Ga > In$.[22]

Since thallium is similar to alkali metals, it has the same behavior as cesium, which has been shown to accumulate in tumors.[23] Tl-201 was used as an imaging agent in 173 malignant tumors and 76 benign lesions.[24] The sensitivity, specificity, and accuracy were 0.64, 0.61, and 0.63, respectively. Sensitivity was good in thyroid cancer (0.91) and fair in primary lung cancer (0.70) and primary liver cancer (0.71). Compared with Ga-67, Tl-201 appears to have a higher sensitivity in thyroid cancer and nearly the same sensitivity in primary lung cancer.

In women with breast cancer, the rate of response to antiestrogenic treatment increases with the content of estrogen receptors in the tumor. Gamma-emitting compounds that bind to the receptors in vivo would allow a noninvasive detection of those tumors having estrogen receptors. Gatley et al.[25] have studied the distribution of 17β-(16-[^{125}-I] iodo)-estradiol (I-E$_2$) in tumor bearing and normal rats, and mean tumor-to-blood ratios of 1.4, 5.5, and 8.7 were seen at 1 hour in rats with transplanted, spontaneous and N-nitrosomethylurea-induced tumors, respectively. McElvany et al.[26] also used 16α[Br-77] bromoestradiol successfully at high specific activity to image carcinogen induced mammary tumors in rats in preliminary studies to image breast tumors in patients.

Several investigators have studied deoxyribonucleic acid (DNA) of high molecular weight as a carrier for cytotoxic drugs in cancer therapy because tumor cells may have a greater endocytic activity than normal cells. Radioiodinated heat-denatured DNA and some related nucleic acids localized in tumor tissues. Retention of radioactivity by tumors was longer than for other organs and localization of tumor tissues was possible by scintigraphic technique 8 to 24 hours after intravenous injection.[27]

External gamma-camera imaging and tissue radioassay were used to study regional lymph node localization of Tc-99m-labeled liposomes in Rd/3 tumor bearing rats following interstitial injection.[28] Radiolabeled liposomes may be useful for diagnosis of lymphatic spread of tumor and may be of value as an in vivo test of immunological function in regional lymph nodes.

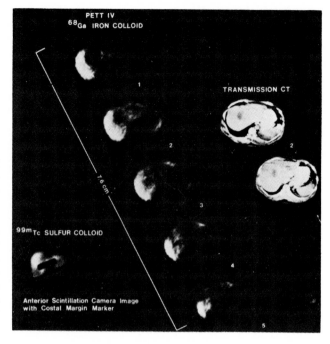

Figure 5. 57-year-old man with hepatic metastases of colon carcinoma. Anterior image with Tc-99m sulfur colloid (lower left) shows multiple focal defects in the liver. Lesions are readily identified on PET images (middle) using Ga-68 iron colloid and correspond with CT findings (right). *(From Am J Roentgenol 1981; 136:687, with permission.)*

POSITRON EMISSION TRANSAXIAL TOMOGRAPHY

The noninvasive measurement of regional tissue function by radiographic techniques demands that the isotope introduced into the body both trace the physiological pathway of interest and emit gamma rays. The monitoring of these rays as they emerge from the tissues, with radiation detectors placed external to the body, allows the local content of tracers to be measured. The combination of certain positron-emitting radionuclides and positron emission computed axial tomography offers both these conditions. The detection of the paired annihilation photons, resulting from capture of emitted positrons, affords a means for recording emission tomograms. In PET the spatial resolution is independent of depth, and absolute correction for tissue absorption is possible.

PET (Fig. 5) offers the opportunity to examine biochemical events as they are actually occurring without having to physically disrupt the tissue. One can thereby view discrete metabolic processes associated with neoplastic development in order to demonstrate in-situ the effectiveness of therapy and the appropriate treatment for related complications. By injecting a particular biochemical substance, its pathway can be followed through the tissues to better determine the nature of the pathological process. This allows examination in detail of quantitative cellular metabolism and tumor kinetics insitu undisturbed by external manipulation, to determine the effect of the radiotherapy, chemotherapy, and immunotherapy on the cellular behavior of the tumor as it is occurring in vivo. Tumor therapy can be individualized for each patient depending upon the responses observed with PET analysis.

Rapid uptake of F-18 fluoro-deoxyglucose (FDG) was observed in a

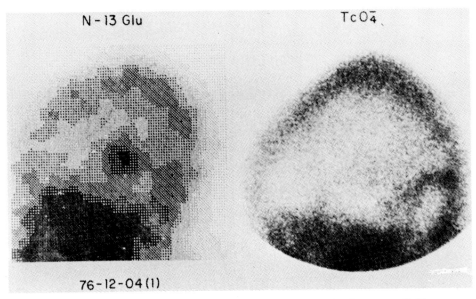

N - 13 Glu   TcO₄⁻

76-12-04 (1)

Figure 6. Left lateral image of head with L-(N-13) glutamate (left) shows a localization of cerebral neuroblastoma. Pertechnetate image (right) shows minimal abnormal uptake of radioactivity. *(From J Nucl Med 1982; 23:685, with permission.)*

variety of transplanted and spontaneous tumors in animals.[29] Tumor-to-normal tissue and tumor-to-blood ratios ranged from 2.10:9.15 and 2.61:17.82, respectively. Uptake of [H-3] 2-deoxyglucose was also studied in BALB/c mice with EMT-6-sarcoma, in buffalo rats with Morris 7777 hepatoma, and in 8 dogs with spontaneous tumors.[30] High tumor-to-tissue ratios were observed for all tumor types, and the studies provide additional support for the concept that PET can be used to obtain functional images of important metabolic processes of tumors, including glycolysis. Five patients who had undergone radiation therapy for cerebral tumors, and whose conditions were deteriorating, were examined by means of PET with F-18 FDG.[31] Two cases of radiation necrosis were distinguished from the three recurrent tumors as the rate of glucose utilization in the necrosis was markedly reduced compared with viable tumor and normal brain parenchyma. In recurrent gliomas, the glucose metabolic rates was actually elevated over normal tissue.

PET with C-11 tryptophan is capable of defining both morphological and functional alterations in the pancreas.[32] Tumors as small as 2 cm in diameter can be detected, and C-11 tryptophan images are technically superior to Se-75 selenomethionine images.

Enzymatically synthesized L-glutamic acid labeled with N-13 (Fig. 6) has been investigated for its potential value as an imaging agent for tumors involving bone and joints.[33] Eleven patients with untreated primary Ewing's sarcoma were studied, and seven were repeatedly scanned during chemotherapy using N-13 L-glutamate and Tc-99m MDP. The untreated primary tumor was distinctly visualized in all cases (Fig. 7). N-13 uptake often decreased more markedly than Tc-99m uptake during chemotherapy. Uptake

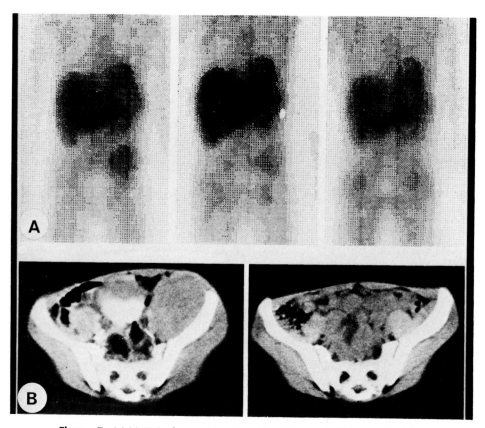

Figure 7. (a) N-13-L glutamate scans of an 8-year-old boy undergoing chemotherapy. Pertechnetate (left) and posttreatment (center and right) scans show decreasing N-13 uptake in Ewing's sarcoma arising from anterior aspect of left iliac bone. (b) Pre-(left) and post-(right) treatment CT images show a reduction of soft tissue mass size. *(From Radiol 1982; 142:497, with permission.)*

of [H-3] thymidine was studied in mice with sarcoma, in rats with hepatoma, and in dogs with spontaneous tumors.[34] High tumor-to-tissue ratios were observed for all tumor types assayed.

Higher tumor-to-nontumor concentration ratios were generally given by 1-aminocyclobutane [C-11] carboxylic acid than [C-14] 1-aminocyclopentane-carboxylic acid.[35] In applying the present technology of PET, the following constraints have to be recognized: (1) the short half-lives of radioisotopes demand that these isotopes be produced at the site of application and by the cyclotron, (2) a steady state level of tracer is required within the tissues during the time of the scan, and (3) the physiological fate of the administered tracers must be known to extract meaningful functional data.

NUCLEAR MAGNETIC RESONANCE IMAGING

NMR imaging is based on the magnetic properties of protons, which spin and act as magnetic dipoles. Placed in a strong magnetic field, they tend to align, producing a net magnetic vector, or moment, oriented parallel to the direction

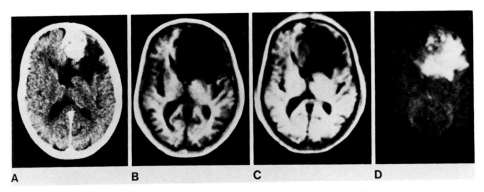

Figure 8. CT (**a**) and NMR head images with different pulse sequence (**b,c,d,**) show a metastatic lesion of nasopharyngeal carcinoma associated with edema. *(From Am J Neuroradiol 1982; 3:474, with permission.)*

of the imposed magnetic field. Application of a radiofrequency (RF) pulse of a specific frequency displaces the net magnetic moment by an amount determined by the strength and duration of the pulse. This frequency is directly proportional to the strength of the magnetic field. After the pulse, the protons can emit an RF signal as they return to their original orientation. The frequency of the signals emitted by protons after irradiation with an RF pulse will reflect their position.

Since the concept of tumor detection by NMR imaging was introduced in 1971,[36] the diagnostic potential of this modality in the discrimination between benign and malignant disease has been viewed with interest. NMR T_1 and T_2 relaxation times have been used to distinguish normal from malignant breast tissue in vitro,[37] and NMR imaging by the FONAR method exhibited lowered proton density than normal tissue.[38] Normal breasts and breasts with extensive fatty replacement were found to have the lowest T_1 values, whereas T_1 values of malignant tissue were elevated.

While malignant brain tumors (Fig. 8) display the same basic changes as benign tumors, there is frequently much more associated edema. Within the tumor, discernment of more internal structure than with x-ray computed tomography including necrosis, cyst formation, and hemorrhage, frequently is possible. The T_1 values of benign brain tumors were generally shorter than those of malignant tumors. A possible metastasis was seen in the brainstem, although this was not demonstrated with CT.[39]

Because fluid and fat have opposite NMR properties, NMR (Fig. 9) is accurate in distinguishing sinus lipomatosis and peripelvic cysts.[40] It is superior to CT, where partial volume averaging sometimes becomes a problem.

Proton NMR was used in a study of 40 patients with thyroid tumors following thryoidectomy.[41] Increased T_1 and T_2 were observed for benign cold nodules, an increase in T_1 alone for nodules with increased uptake, and a wide fluctuation in T_1 and T_2 for multinodular goiters. The four cancers did not show a distinctive proton NMR pattern in comparison with the other nodular structures studied.

All preliminary studies suggest that developing NMR techniques may potentially play a useful role in one or more of the following areas: (1)

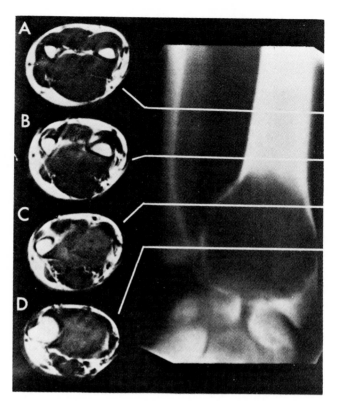

Figure 9. Serial NMR images of distal forearm and corresponding tomogram show a slight decrease in signal intensity in the medullary cavity of radius at level (**b**), consistent with proximal extent of giant cell tumor. The medullary cavity is expanded at levels (**c**) and (**d**) with a marked inhomogenous reduction in NMR signal, consistent with tumor involvement on the radiograph (right).

diagnosing cancer in a biopsy tissue specimen, (2) diagnosing the presence of malignant tissue in the whole human host, (3) detecting the location of malignancy in the human host to aid in directing specific therapy to the tumor while minimizing host toxicity, (4) monitoring the growth or regression of malignant tumors while undergoing treatment, and (5) allowing for a new chemical classification of malignant tumors.

NMR technology is being refined and sensitivity and specificity evaluated. As NMR data becomes more reproducible, a major effort at indexing values for all normal and malignant tissues will be needed. Emphasis on providing this technology at a reasonable cost also will be important if these exciting advances are to become generally available in clinical medicine.

REFERENCES

1. Goldenberg DM: Oncofetal and other tumor-associated antigens of the human digestive system. Curr Top Pathol 1976; 63:289–342
2. Pressman D: The development and use of radiolabeled antitumor antibodies. Cancer Res 1980; 40:2960–2964
3. McCardle RJ, Harper PV, Spar IL, et al: Studies with iodine-131-labeled antibody to human fibrinogen for diagnosis and therapy for tumors. J Nucl Med 1966; 7:837–847
4. Spar I, Goodland RL, Desiderio MA: Immunological removal of circulating ^{131}I-labeled rabbit antibody to rat fibrinogen in normal and tumor-bearing rats. J Nucl Med 1964; 5:428–434

5. Krohn KA, Sherman K, Welch MJ: Studies of radioiodinated fibrinogen. I. Physiochemical properties of the ICI, chloramine-T, and electrolytic reaction products. Biochem Biophys Acta 1972; 285:404–409
6. Goldenberg DM, Kim EE, Deland FH, et al: Radioimmunodetection of cancer with radioactive antibodies to carcinoembryonic antigen. Cancer Res 1980; 40:2984–2992
7. Dykes PW, Hine KR, Bradwell AR, et al: Localization of tumor deposit by external scanning after injection of radiolabeled anticarcinoembryonic antigen. Br Med J 1980; 26:220–222
8. Kim EE, Deland FH, Nelson MO, et al: Radioimmunodetection of cancer with radiolabeled antibodies to α-fetoprotein. Cancer Res 1980; 40:3008–3012
9. McManus LM, Naughton MA, Martinez-Hernandez M: Human chorionic gonadotropin in human neoplastic cells, Cancer Res 1976; 36:3476–3481
10. Goldenberg DM, Kim EE, Deland FH, et al: Clinical radioimmunodetection of cancer with radioactive antibodies to human chorionic gonadotropin. Science 1980; 208:1284–1286
11. Ballow B, Levine G, Hakula TR, et al: Tumor location detected with radioactively labeled monoclonal antibody and external scintigraphy. Science 1979; 206:844–846
12. Levine G, Ballou B, Reiland J, et al: Localization of I-131 labeled tumor-specific monoclonal antibody in the tumor-bearing BALB/c mouse. J Nucl Med 1980; 21:570–573
13. Spiegelberg HL, Weigle WO: The catabolism of homologous and heterologous gamma globulin fragments. J Exp Med 1965; 121:323–338
14. Wilbanks, T, Peterson JA, Miller S, et al: Localization of mammary tumors in vivo with ^{131}I-labeled Fab fragments of antibodies against mouse mammary epithelial antigens. Cancer 1981; 48:1768–1775
15. Poulouse KP, Watkins AE, Reba RC, et al: Cobalt-labeled bleomycin a new radiopharmaceutical for tumor localization. A comparative clinical evaluation with gallium citrate. J Nucl Med 1975; 16:839–841
16. Lilien DL, Jones SE, O'Mara RE, et al: A clinical evaluation of In-111 bleomycin as a tumor imaging agent. Cancer 1975; 35:1036–1041
17. Goodwin DA, Meares CF, De Riemer LH, et al: Clinical studies with In-111 BLEDTA, a tumor-imaging conjugate of bleomycin with a bifunctional chelating agent. J Nucl Med 1981; 22:787–792
18. Mori T, Odori T, Sakamoto T, et al: Clinical results of ^{99m}Tc-labeled bleomycin scintigraphy for tumor imaging, in Proceedings of the First World Congress of Nuclear Medicine, Tokyo, 1974, p 703
19. Breslow K, Halpern SE, Schwartz FC, et al: Stability studies and tumor uptake of a technetium-tetracycline complex. J Nucl Med 1974; 15:987–990
20. Young D, Vine E, Ghanbarpour A, et al: Metabolic and distribution studies with radiolabeled 5-fluorouracil. Nukleamedizin 1982; 21:1–7
21. Hisada K, Suzuki Y, Hiraki T, et al: Clinical evaluation of tumor scanning with ^{169}Yb-citrate. Radiol 1975; 116:389–393
22. Hart MM, Adamson RH: Antitumor activity and toxicity of salts of inorganic group IIIa metals: Aluminum, gallium, indium, and thallium. Proc Nat Acad Sci 1971; 68:1623–1626
23. Nishiyama H, Sodd VJ, August L, et al: Tumor scanning agent: Reappraisal of cesium, ^{129}C$_5$CL. J Nucl Med 1973; 14:635
24. Hisada K, Tonami N, Miyamae T, et al: Clinical evaluation of tumor imaging with ^{201}Tl chloride. Radiol 1978; 129:497–500
25. Gatley SJ, Shaughnessy WJ, Inhorn L, et al: Studies with 17β-(16-[^{125}I]iodo)-estradiol, an estrogen-receptor-binding radiopharmaceutical, in rats bearing mammary tumors. J Nucl Med 1981; 22:459–464

26. McElvany KD, Katzenellenbogen JA, Shafer KE, et al: 16α-Br-77 bromoestradiol: Dosimetry and preliminary clinical studies. J Nucl Med 1982; 23:425–430

27. Ithakissios D, Hapke B: Radioiodinated DNA as potential tumor-imaging agent. J Nucl Med 1979; 20:785–788

28. Osborne MP, Richardson VJ, Jeyasingh K, et al: Potential applications of radionuclide-labeled liposomes in the detection of the lymphatic spread of cancer. Int J Nucl Med Biol 1982; 9:47–51

29. Som P, Atkins HL, Bandoypadhyay D, et al: A fluorinated glucose analog. 2-fluoro-2-deoxy-D-glucose (F-18): Nontoxic tracer for rapid tumor detection. J Nucl Med 1980; 21:670–675

30. Larson SM, Weiden PL, Grumbaum Z, et al: Positron imaging feasibility studies II: Characteristics of 2-deoxyglucose uptake in rodent and canine neoplasms. J Nucl Med 1981; 22:875–879

31. Patronas NJ, Dichiro G, Brooks RA, et al: [18F] flurodeoxyglucose and positron emission tomography in the evaluation of radiation necrosis of the brain. Radiol 1982; 144:885–889

32. Kirchner PT, Ryan J, Zalutsky M, et al: Positron emission tomography for the evaluation of pancreatic disease. Semin Nucl Med 1980; 10:374–391

33. Reiman RE, Rosen G, Gelbard AS, et al: Imaging of primary Ewing's sarcoma with 13N-L-glutamate. Radiol 1982; 142:495–500

34. Larson SM, Weiden PL, Grunbaum Z, et al: Positron imaging feasibility studies. I. Characteristics of [3H] thymidine uptake in rodent and canine neoplasms. J Nucl Med 1981; 22:869–874

35. Washburn LC, Sun TT, Byrd BL, et al: 1-aminocyclobutane I 11CI carboxylic acid, a potential tumor-seeking agent. J Nucl Med 1979; 20:1055–1061

36. Damadian R: Tumor detection by nuclear magnetic resonance. Science 1971; 171:1151–1153

37. Bovée WM, Getreuer KW, Schmidt J, et al: Nuclear magnetic resonance and detection of human breast tumor. J Natl Cancer Inst 1978; 67:53–55

38. Ross RJ, Thompson JS, Kim K, et al: Nuclear magnetic resonance imaging and evaluation of human breast tissue: Preliminary clinical trials. Radiol 1982; 143:195–205

39. Bydder GM, Steiner RE, Young IR, et al: Clinical NMR imaging of the brain: 140 cases. Am J Neuro Radiol 1982; 3:459–480

40. Hricak H, Crooks L, Sheldon P, et al: Nuclear magnetic resonance imaging of the kidney. Radiol 1983; 146:425–432

41. de Certaines J, Herry JY, Lancien G, et al: Evaluation of human thyroid tumors by proton nuclear magnetic resonance. J Nucl Med 1982; 23:48–51

APPENDIX 1

Review of Nuclear Imaging Procedures in Oncology

I. Head and Neck Imaging
  A. Brain Scan and Flow Study
    1. Indication
      Identifying:
      (a) Primary or secondary neoplasm.
      (b) Cerebral infarction, hematoma, and contusion.
      (c) Subdural hematoma.
      (d) Aneurysm or arteriovenous malformation.
      (e) Meningoencephalitis.
      (f) Cerebral abscess.
      (g) Brain death.
    2. Preparation of the patient
      None.
    3. Instrumentation
      Gamma camera; general all-purpose collimator (GAP); computer.
    4. Radiopharmaceutical
      Tc-99m diethylene triamine pentacetic acid (DTPA) or glucoheptonate (GH). Adult dose 15-20 mCi.
    5. Method (Fig. 1)
      (a) Anterior dynamic flow study is usually performed with the patient supine and the camera positioned in the forehead-nasion position. Posterior or vertex flow study may be indicated in certain clinical settings.
      (b) Using a bolus intravenous injection, rapid sequential scintiphotos at 1 to 2 second intervals are taken.
      (c) Without moving the patient immediate blood-pool imaging is obtained with 300 to 500 K counts in the position used for perfusion and each lateral position.

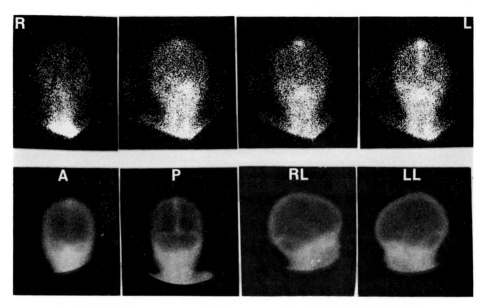

Figure 1. Routine anterior dynamic study (upper row) and static images (lower row) of the head using Tc-99m DTPA show normal symmetrical activities and no focal or generalized abnormal activities.

   (d) Anterior, posterior, each lateral, and vertex static images
       should be obtained at 1 to 3 hours.
B. Cisternography
   1. Indication
       (a) To confirm the presence and differential diagnosis of hydro-
           cephalus.
       (b) To confirm the presence of porencephalic cyst.
       (c) To confirm the presence of cerebrospinal fluid (CSF) leak.
   2. Preparation of the patient
      Informed consent required for the lumbar puncture.
   3. Instrumentation
      Gamma camera; medium energy collimator.
   4. Radiopharmaceutical
      In-111 DTPA or Yb-169 DTPA. Adult dose 0.5 to 1 mCi
   5. Method
   (a) Under usual sterile precautions, the standard lumbar puncture
       is performed. Special attention must be given so that very little
       or no CSF is removed, even with the measurement of CSF
       pressure.
   (b) The radioactive material in the syringe is gently mixed with
       the CSF, and the total mixture is then reinjected into the
       lumbar subarachnoid space.
   (c) Immediate images of the lumbar area are obtained, and the
       imaging of head is obtained in 2 to 4 hours in the anterior,
       either lateral, and vertex positions. These studies are repeated
       at 24 hours and may be repeated at 48 hours under abnormal
       conditions.

(d) Special note: If the spinal tap is bloody, do not inject the radiopharmaceutical. Withdraw the needle and reattempt study the next day.

C. CSF Shunt Study
  1. Indication
  To evaluate the patency of ventriculo-peritoneal, ventriculoatrial and subarachnoid-peritoneal shunts.
  2. Preparation of the patient
    (a) Informed consent required; sedation if necessary.
    (b) Razor, betadine, lumbar puncture tray.
  3. Instrumentation
  Gamma camera; GAP or medium energy collimator.
  4. Radiopharmaceutical
  1-2 mCi Tc-99m DTPA or 0.5 mCi in 111 DTPA.
  5. Method
    (a) With sterile precaution, the radiopharmaceutical is injected into the site of choice (usually reservoir of the shunt by neurosurgeon).
    (b) Imaging begins immediately and continued at certain intervals to follow the radioactivity flow.

D. Dacroscintigraphy
  1. Indication
  To evaluate the patency of the nasolacrimal duct, especially postradiation.
  2. Preparation of the patient
  None.
  3. Instrumentation
  Gamma camera; pinhole collimator; computer; special head-rest chair; micropipette.
  4. Radiopharmaceutical
  Tc-99m sodium pertechnetate.
  5–10 mCi in each eye.
  5. Method
    (a) The patient sits with forehead-nasion against the collimator.
    (b) One drop of the radiopharmaceutical is placed in the medial canthus of each eye and imaging is started immediately. Blotting may be required, using cotton applicator.
    (c) Obtain 5 second images for 1 to 2 minutes, then every 15 seconds for 10 minutes for computer program.

E. Eye Tumor Localization
  1. Indication
  To localize in   :ular masses and to differentiate benign from malignant lesio.   .nelanoma).
  2. Preparation of th patient
  None.
  3. Instrumentation
  Side window ocular probe (gas sterilization) scaler.
  4. Radiopharmaceutical
  P-32 sodium phosphate 600–800 uCi.

5. Method
    (a) The patient is given the radiopharmaceutical orally in 1 to 2 ounces of water.
    (b) At 72 to 96 hours the patient is brought to the operating room for eye surgery. Under sterile precautions, the probe is placed on the external surface of the globe over the lesion and over a separate control area.
    (c) A minimum of 5 1-minute counts is obtained and averaged. Melanoma usually contains two or more times as many counts as control area.

F. Salivary Gland Scan
    1. Indication
        (a) To evaluate salivary gland tumors (especially Warthin's tumor) or other pathology.
        (b) To determine function of salivary glands.
    2. Preparation of the patient
    None.
    3. Instrumentation
    Gamma camera; GAP collimator; computer.
    4. Radiopharmaceutical
    Tc-99m sodium pertechnetate 5–10 mCi.
    5. Method
        (a) With the camera centered over the patient's nose in the frontal position or the torcular area in the posterior position, the radiopharmaceutical is injected intravenously by the bolus method, and the 4-second scintiphotos for nine exposures are obtained.
        (b) 3-minute static imagings are followed for 10 to 20 minutes, and the camera is then centered over each zygomatic arch, and the images are made.
        (c) Additional images may be obtained using the pinhole collimator or following the use of lemon to promote emptying of salivary glands.

G. Thyroid Perfusion and Scan
    1. Indication
        (a) To determine the presence and functional state of thyroid neoplasia.
        (b) To evaluate patients with abnormal thyroid function or multinodular goiter.
        (c) To evaluate the presence of lingual, or substernal thyroid mass.
        (d) To determine size and shape of thyroid.
        (e) To determine recurrent or metastatic functional thyroid cancer.
        (f) To evaluate vascularity or hemorrhage in the nodule greater than 2 cm in size.
    2. Preparation of the patient
    The patient should not have had a thyroid blocking agent (iodine contrast for at least 2 weeks and thyroxin for 6 weeks).

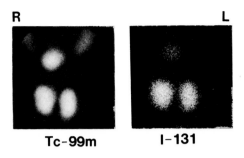

R   **L**

Tc-99m   I-131

Figure 2. Anterior static images of the neck using Tc-99m pertechnetate and I-131 show a uniform uptake of radioactivity in right and left lobes of the thyroid. Note a functioning sublingual thyroid in the midline.

3. Instrumentation

Gamma camera; low-energy conveying collimator for perfusion; pinhole collimator for static images.

4. Radiopharmaceutical

5 to 10 mCi Tc-99m sodium pertechnetate.

100 to 200 uCi Na-123-I.

50 to 100 uCi Na-131-I.

5. Method

(a) In patients with a large solitary nodule, a perfusion study with Tc-99m pertechnetate is usually performed. The patient lies supine, and the camera is centered over the thyroid area. Using the intravenous bolus injection, nine 2-second scintiphotos are obtained, and the immediate blood-pool image (300 K counts) is then made.

(b) Approximately 30 minutes following the injection of Tc-99m the anterior and each oblique static images of the thyroid (Fig. 2) are obtained. For I-131 scan, 24-hour image is routinely made. For I-123 scan, 4 to 6 hour image is usually obtained. Additional image of the midchest may be made if needed.

II. Chest Imaging

A. Lung Ventilation/Perfusion ($\dot{V}/\dot{Q}$) Scans

1. Indication

(a) Evaluation of lung cancer for both the involved and uninvolved lung, pre- and postradiation for surgery (resectability and operability).

(b) Evaluation of bronchial obstruction or bronchopleural fistula.

(c) Diagnosis and management of pulmonary emboli.

(d) Evaluation of regional ventilation and perfusion in emphysema.

(e) Study of pulmonary venous hypertension.

(f) Quantitation of right to left cardiac or intrapulmonary shunt.

2. Preparation of the patient

None.

3. Instrumentation

Gamma camera; GAP collimator; xenon-ventilation unit or aerosol nebulizer, computer.

4. Radiopharmaceutical

Xe-133 gas or Xe-133 dissolved in saline: 10 to 20 mCi.

Tc-99m DTPA aerosol: 10 to 15 mCi.

Tc-99m HAM or MAA: 2 to 5 mCi.

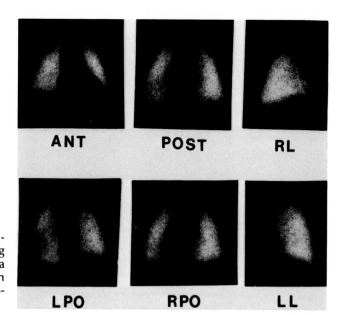

Figure 3. Routine multiple images of the lung with Tc-99m MAA shows a homogenous activity in both lungs. Mild cardiomegaly is noted.

5. Method
   (a) The order of usual V̇/Q̇ scan is Xe-133 inhalation followed by Tc-99m perfusion, but Tc-99m aerosol study is performed following perfusion study.
   (b) Ventilation study is performed in the posterior position with the patient sitting erect if possible; otherwise, posterior supine.
   (c) After Xe-133 gas inhaled through a closed system, an initial 20-second single breath image is obtained followed by rebreathing until equilibrium occurs (usually 3 to 5 minutes). Then, 60-second washout images are made while the patient exhales the gas. This should be carried out until the lungs are free of radioactivity (usually 3 to 10 minutes).
   (d) Except under unusual circumstances, Tc-99m particles are injected with the patient horizontal during normal breathing. Anterior, posterior, each lateral, and each oblique views are then obtained (Fig. 3).
   (e) To generate regional V̇/Q̇ ratio by the computer, the patient should not be moved between the start of ventilation study until after the first posterior image is obtained.
   (f) Additional brain and renal images are made in cases of shunt evaluation by the computer.
B. Ventricular Function Test with or without Stress
   1. Indication
      (a) Evaluation of certain drug effect on ventricular function (cardiotoxicity).
      (b) Evaluation of ventricular ejection fraction and wall motion for pre- and post-bypass operation and postinfarction.

       (c) Study of intracardiac shunt, mean pulmonary transit time, and cardiac output.

  2. Preparation of the patient

    None; patients usually require adequate hydration and monitoring.

  3. Instrumentation

    Gamma camera; low-energy high-resolution collimator; ECG gating and monitoring unit; stress equipment; computer.

  4. Radiopharmaceutical

    In vivo or vitro labeled Tc-99m RBC 20 mCi.

  5. Method

    (a) The patient is intravenously given 7 to 10 mg unlabeled pyrophosphate approximately 30 minutes prior to the study.

    (b) The patient is placed in the supine or semierect position on the stress table.

    (c) Following the bolus injection of 20 mCi, Tc-99m sodium pertechnetate, 0.5-second serial anterior scintiphotos for 60 seconds of the first pass of the radiopharmaceutical through the heart are obtained, and the data is accumulated by the computer in systole and diastole for about 5 minutes after.

    (d) With appropriate ECG monitoring, the patient bicycles to the maximum obtainable stress that can be maintained for at least 60 seconds.

    (e) Acquisition of ventricular function in 45° left anterior oblique (LAO) position is accomplished during the 60 seconds.

III. Abdominal Imaging

  A. Liver-Spleen Scan

    1. Indication

      (a) Preoperative evaluation of hepatic metastases from known malignancies.

      (b) Detection of focal space-occupying lesions such as tumor, cyst, and abscess.

      (c) Evaluation of abdominal mass.

      (d) Follow-up of patients with liver metastases undergoing various treatments.

      (e) Evaluation of the size, shape, and position of liver or spleen.

      (f) Workup of patients with diffuse liver disease such as cirrhosis or hepatitis.

      (g) Localization of hepatic lesions for biopsy or drainage.

      (h) Detection and follow-up of liver or spleen contusion or hematoma.

      (i) Evaluation of vascularity in space-occupying lesion.

    2. Preparation of the patient

      None.

    3. Instrumentation

      Gamma camera; GAP collimator.

    4. Radiopharmaceutical

      3-5 mCi Tc-99m sulfur colloid.

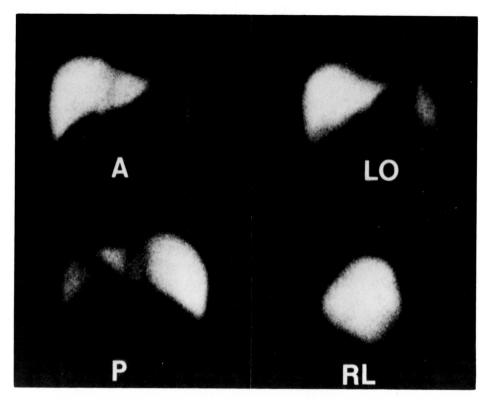

Figure 4. Routine static images of the upper abdomen using Tc-99m sulfur colloid show normal sized liver and spleen with uniform activity.

5. Method ·
    (a) Using the bolus injection, nine rapid serial images at 4-second intervals are obtained for perfusion study.
    (b) Approximately 10 to 15 minutes following the injection, static anterior, posterior, each lateral, and occasional oblique images of upper abdomen are made (Fig. 4).
B. Hepatobiliary Scan
  1. Indication
    (a) Evaluation of biliary duct obstruction or atresia.
    (b) Evaluation of gallbladder disease (acute cholecystitis) or function.
    (c) Postop or traumatic bile leakage or biliary cutaneous fistula.
    (d) Detection of choledochal cyst.
    (e) Evaluation of postop bile reflux into the stomach.
    (f) Evaluation of hepatocytic function and bile flow rate.
  2. Preparation of the patient
    Patient fasts for a minimum of 2 hours.
  3. Instrumentation
    Gamma camera; GAP collimator.

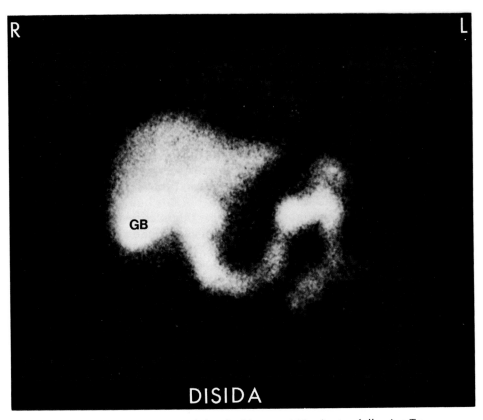

Figure 5. Anterior upper abdominal image at 30 minutes following Tc-99m DISIDA injection shows a visualization of gallbladder (GB) and the activity excreted into the proximal jejunum.

    4. Radiopharmaceutical

       Tc-99m diisopropyl iminodiacetic acid (DISIDA) 5 to 8 mCi.

    5. Method

      (a) 500 K anterior image of the liver is obtained at 5 minutes following the intravenous injection of radiopharmaceutical, then every 15 minutes until small bowel is visualized (Fig. 5). A right lateral image of the liver is made at 30 minutes to confirm the visualization of gallbladder.

      (b) If the gallbladder activity is visualized within 2 hours without emptying into the small bowel, a fatty meal or CCK may be administered to discern gallbladder contraction.

      (c) In case of suspicious biliary atresia, a 24-hour image may also be indicated.

C. Gastrointestinal Bleeding Study

    1. Indication

      (a) Melena.

      (b) Hematemesis.

      (c) Meckel's diverticulum.

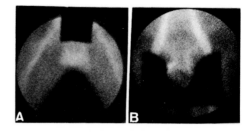

Figure 6. Anterior static images of the testes immediately following Tc-99m pertechnetate injection show a uniform activity in normal testes (**a**) and an irregular decreased activity in left testis with seminoma (**b**).

Overnight fasting.
3. Instrumentation
Gamma camera; medium-energy collimator.
4. Radiopharmaceutical
250 uCi Se-75 selenomethionine.
5. Method
(a) Wait 30 minutes for gastric emptying following the ingestion of 60 ml lipomul, and then inject the radiopharmaceutical.
(b) Patient is placed supine, and the serial 10 minute anterior images are obtained for 1 to 2 hours.
H. Testicular Perfusion and Imaging
1. Indication
(a) To evaluate testicular mass (tumor or abscess).
(b) To detect testicular torsion, epididymorchitis or varix.
2. Preparation of the patient
None.
3. Instrumentation
Gamma camera; GAP collimator.
4. Radiopharmaceutical
10 mCi Tc-99m sodium pertechnetate or Tc-99m DTPA.
5 mCi Tc-99m sulfur colloid.
5. Method
(a) Following the intravenous injection of radiopharmaceutical nine 4-second images of testes are obtained.
(b) Immediate static images (500 K counts) (Fig. 6), followed by 5- and 10-minute images.
IV. Vascular Structure Imaging
A. Arterial Catheter Flow Study
1. Indication
(a) To evaluate the perfusions of supplying organ or tumor.
(b) To guide adequate division of chemotherapeutic agents into bilateral internal iliac arteries.
2. Preparation
None.
3. Instrumentation
Gamma camera; GAP collimator.
4. Radiopharmaceutical
2 mCi Tc-99m MAA in 0.5 to 1.0 ml.

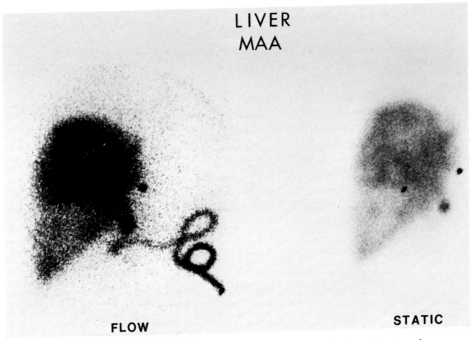

LIVER
MAA

FLOW

STATIC

Figure 7. Anterior images of the upper abdomen during (flow) and after (static) the infusion of Tc-99m MAA show only perfusion of MAA to the right lobe of the liver. No extrahepatic activity is noted.

5. Method (Fig. 7)
   (a) Introduce the radiopharmaceutical into the venotubing of chemotherapy infusion pump without changing flow rate.
   (b) Obtain 500 K static image when the radioactivity fills arterial catheter.
   (c) Another anterior (and each lateral for pelvic case) static image when the activity clears from the catheter.
   (d) Additional lung image to evaluate arteriovenous shunt in the tumor bed.
B. Venous Flow Study
   1. Indication
      (a) To evaluate patency of deep venous system.
      (b) To evaluate organized venous thrombosis and collateral circu- lation.
   2. Preparation of the patient
      None.
   3. Instrumentation
      Gamma camera; GAP collimator.
   4. Radiopharmaceutical
      2–4 mCi Tc-99m HAM or MAA for lower extremity.
      2–4 mCi Tc-99m sulfur colloid or MAA for upper extremity.
C. Lymphoscintigraphy
   1. Indication

(a) To evaluate lymphatic flow and lymph nodes draining organ or region involved with a malignant tumor.

(b) To evaluate congenital or acquired lymphedema.

2. Preparation of the patient

None, except for informed consent.

3. Instrumentation

Gamma camera; GAP and high-resolution collimators.

4. Radiopharmaceutical

1–2 mCi Tc-99m antimony colloid 0.5 ml.

5. Method

(a) The anatomic site containing the lymphatics leading to potentially involved lymph nodes is injected.

(b) Anterior and oblique views are obtained every 30 minutes for 2 to 6 hours, depending on lymphatic flow.

V. Whole-Body Imaging

A. Bone Scan

1. Indication

(a) To screen preoperatively (staging) patients with cancers known to metastasize early to bone.

(b) To determine precise localization for biopsy and extent of bone metastases (for radiotherapy plan).

(c) To evaluate clinically suspicious or radiographically difficult areas of bone metastases.

(d) To monitor the effects of various treatments.

(e) To evaluate stress or traumatic fracture.

(f) To diagnose osteomyelitis.

2. Preparation of the patient

Reasonably good hydration; catheterization if patient is unable to void.

3. Instrumentation

Gamma camera: GAP collimator.

4. Radiopharmaceutical

15-20 mCi Tc-99m MDP.

5. Method

(a) For tumors, anterior and posterior static images of whole skeleton are obtained 2 to 3 hours following the injection (Fig. 8).

(b) For osteomyelitis, dynamic flow study and immediate blood-pool image are obtained in addition to delayed (2 to 3 hour) static image for the affected area and its contralateral areas.

B. Bone Marrow Scan

1. Indication

(a) Evaluation of general hypoplasia of the RES system in patients with myelofibrosis, leukemia, multiple myeloma, and disseminated carcinomatosis.

(b) Detection of focal diseases in bone marrow, such as plasmacytoma, or Hodgkin's disease.

(c) Evaluation of RES hyperplasia in patients with hemolytic anemia and polycythemia.

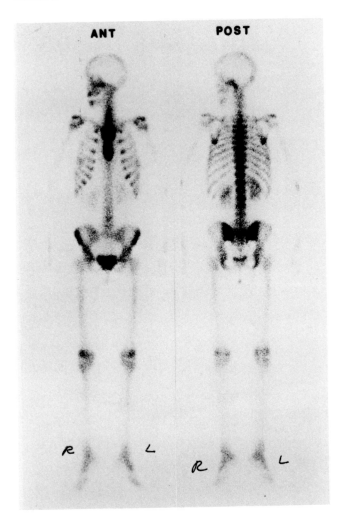

Figure 8. Anterior and posterior static images of the whole body with Tc-99m MDP show no focal area of abnormally increased or decreased activity in the bone.

    (d) Evaluation of RES response to chemotherapy or irradiation.
2. Preparation of the patient
    None.
3. Instrumentation
    Gamma camera; medium-energy or GAP collimator.
4. Radiopharmaceutical
    2–3 mCi In-111 Cl or transferrin.
    10–15 mCi Tc-99m sulfur colloid.
5. Method (Fig. 9)
    (a) Anterior and posterior images of the total skeleton are obtained at 24 to 72 hours postinjection of In-111 Cl.
    (b) To evaluate malignant madiastinal tumors such as lung cancer, seminoma, or undifferentiated thyroid cancer.
    (c) To differentiate hepatoma from cirrhotic nodule.
    (d) To determine the presence of pyogenic abscess.

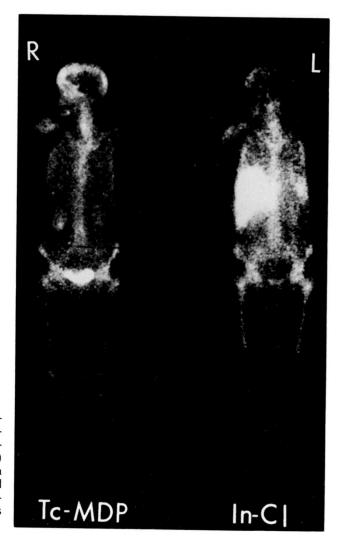

Figure 9. Anterior whole-body images of bone (Tc-99m MDP) and bone marrow (In-111 chloride) scans show no focal area of abnormally decreased activity. Normal distribution of marrow activity is seen.

C. Gallium Scan
   1. Indication
      (a) To aid in the staging of lymphoma (especially Hodgkin's disease).
      (b) To evaluate malignant mediastinal tumors such as lung cancer, seminoma, or undifferentiated thyroid cancer.
      (c) To differentiate hepatoma from cirrhotic nodule.
      (d) To determine the presence of pyogenic abscess.
   2. Preparation
      (a) Reasonable hydration.
      (b) Bowel cleansing may be required for abdominal imaging if needed.
   3. Instrumentation
      Tomographic (phocon) scanner; computer (for subtraction).

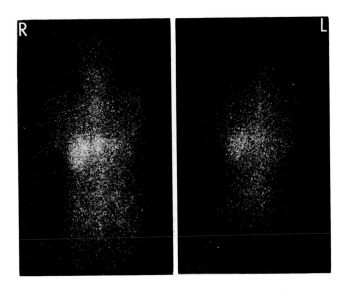

Figure 10. Anterior and posterior whole-body images of Ga-67 citrate scan show no focal area of abnormal accumulation of radiogallium in head, chest, abdomen, and lymphatic channels.

4. Radiopharmaceutical
   Ga-67 citrate.
   10–12 mCi for patients with known tumor.
   5–7 mCi for abscess localization.
5. Method (Fig. 10)
   (a) For tumor localization, anterior, posterior, and lateral views of the appropriate area are obtained at 24 to 48 and 72 hours.
   (b) For detection of abscess, anterior, posterior, and lateral views of suspicious area are obtained at 4 to 6, 24, and 48 hours (if necessary).

APPENDIX 2

Physical Properties of Imaging Radionuclides Mentioned in the Text*

Element	Chemical Symbol (X)	Atomic Number (Z)	Mass Number (A)	Physical Half-Life	Principle Imaging Emission (MeV)*
Americium	Am	95	241	458y	0,060γ Fluorescent
Bromine	Br	35	77	2.38d	0.24γ
Carbon	C	6	11	20.3m	0.97β+
Cobalt	Co	27	55	18.2h	1.50β+
Cobalt	Co	27	57	270d	0.122γ
Chromium	Cr	24	51	27.8d	0.302γ
Fluorine	F	9	18	1.83h	0.635β+
Gallium	Ga	31	67	3.24d	0.93, 0.184, 0.296γ
Gallium	Ga	31	68	1.14h	1.90β+
Gold	Au	79	198	2.70d	0.412γ
Indium	In	49	111	2.81d	0.173, 0.247γ
Indium	In	49	113m	1.66h	0.393γ
Iodine	I	53	123	13.3h	0.159γ
Iodine	I	53	125	60.2d	0.035γ
Iodine	I	53	131	8.05d	0.364γ
Krypton	Kr	36	81m	13s	0.190γ
Mercury	Hg	80	197	2.71d	0.077γ
Nitrogen	N	7	13	9.96m	β+
Oxygen	O	8	15	2.1m	β+
Phosphorus	P	15	32	14.3d	1.710β− Bremstrahlung
Rubidium	Rb	37	82	6.3h	0.78β+
Selenium	Se	34	75	120.4d	0.265, 0.280γ
Technetium	Tc	43	99m	6.0h	0.140γ
Thallium	Tl	81	201	3.08h	0.080 HgK x-ray
Xenon	Xe	54	133	5.27d	0.081γ
Ytterbium	Yb	70	169	31.8d	0.177, 0.198γ
Yttrium	Y	39	90	2.66d	2.27β− Bremstrahlung
Zinc	Zn	30	69m	13.8h	0.439γ

* (from Brucer M Trilinear Chart of the Nuclides. Mallenckrodt Inc., St. Louis, Missouri, 1968.)

APPENDIX 3

Radiopharmaceuticals Used in Common Nuclear Imaging Procedures

Radiopharmaceutical	Procedures
^{99m}Tc-Sodium pertechnetate (NaTcO$_4$)	Thyroid imaging, brain imaging, voiding cystography, testicular imaging, salivary gland imaging, dacroscintigraphy, imaging for Barret's esophagus, imaging for Meckel's diverticulum, gastric mucosa imaging, parathyroid imaging, venogram, and blood-pool imaging.
^{99m}Tc-Sulfur colloid	Liver-spleen imaging, imaging for bleeding, testicular imaging, imaging for esophageal motility, imaging for gastric emptying, imaging for Levine shunt, imaging for transplant rejection, angiogram.
^{99m}Tc-Diisopropyliminodiacetic acid (DISIDA)	Hepatobiliary imaging and gallbladder imaging.
^{99m}Tc-Antimony colloid	Lymphoscintigraphy.
^{99m}Tc-Diethylenetriamine pentacetic acid (DTPA)	Brain imaging, renal blood flow and perfusion, pulmonary ventilation (aerosol).
^{99m}Tc-Human albumin microsphere (HAM) or macroaggregate albumin (MAA)	Pulmonary perfusion imaging, venogram, muscle perfusion imaging, imaging for catheter patency, imaging for Levine shunt patency.
^{99m}Tc-Human serum albumin (HSA)	Blood-pool imaging.
^{99m}Tc-Methylenediphosphate (MDP)	Bone imaging, renal imaging, imaging for intestinal or soft tissue calcification.
^{99m}Tc-Pyrophosphate	Myocardial infarction imaging cerebral infarction imaging, muscle necrosis imaging.

^{123}I or ^{131}I-Sodium iodine	Thyroid imaging.
^{131}I-Hippuran	Renal imaging, imaging for renal tubular function or obstructive uropathy.
^{131}Norcholesterol	Imaging for adrenal cortical lesion.
^{131}I-Iodobenzylquanidine (MIBG)	Imaging for adrenal medullar lesion.
^{111}In-Chloride	Bone marrow imaging.
^{111}In-Diethylene triamine pentacetic acid (DTPA)	Cisternogram, imaging of CSF leak, imaging for VP shunt patency.
^{111}In-Leukocyte	Imaging for inflammation (abscess, osteomyelitis, or nephritis).
^{111}In-Platelet	Imaging for thrombosis, imaging for transplant rejection.
^{67}Ga-Citrate	Imaging for inflammation (abscess, interstitial pneumonitis), imaging for chemotherapy-lung, imaging for tumor (lung cancer, lymphoma, hepatoma or seminoma).

INDEX